ROYAL NAVY UNIFORMS
1930–1945

ROYAL NAVY UNIFORMS
1930–1945

MARTIN J. BRAYLEY

THE CROWOOD PRESS

First published in 2014 by
The Crowood Press Ltd
Ramsbury, Marlborough
Wiltshire SN8 2HR

www.crowood.com

British Library Cataloguing-in-Publication Data
A catalogue record for this book is available from the British Library.

ISBN 978 1 84797 844 8

Dedication
Dedicated to Leading Cook Derrick G. Brayley, D/MX 51983, HMS *Gloucester.* Killed in action 22 May 1941.

Acknowledgements
As with any work of this type, the finished title is the result of the collaboration of many individuals and institutions, without whose assistance it would not have been possible to complete. I wish to thank the following for their generous help in the compilation of this work: Alan Watson, for kindly granting access to the finely preserved World War II MGB *Medusa.* Heather Johnson and Margaret Newman of the Royal Naval Museum Library, Portsmouth, for assisting with research material. Marc Farrance of Explosion, Museum of Naval Firepower, for granting access to the museum's 4.7in AA gun. Tony Drury, of the RN Research Archive. The Historical Diving Society, for access to a DSEA. The Royal Navy Research Archive for kindly providing a number of period images. Richard Noyce of the National Museum of the Royal Navy, Portsmouth. Steve Britton of Britton at War for the loan of period items. Paul Laidlaw for providing images of the Officers' 5B Dress. Reg Woodward, HMS *Mauritius* 1944–6. Lieutenant (A) Keith Remington RNVR for the loan of period images. My partner Jeanette Palubicki for her proof-reading skills and support throughout the project. The guys who kindly agreed to model the uniforms: Matthew Barton, Toby Brayley, Jonathan Davies, Matthew Palubicki, Mike Palubicki, John Price, Robert F. Stedman and Alistair Little. And in no particular order, John Bodsworth, Roger Dennis, David B. Howell, John F. Lamont, Lee Parker, and for his continued support, Canadian historian and author Ed Storey.

Typeset by Jean Cussons Typesetting, Diss, Norfolk

Printed and bound in India by Replika Press Pvt Ltd

Contents

Introduction

> The Officers, Men and Boys of His Majesty's Fleet and the Royal Marines shall wear such uniforms as the Admiralty in pursuance of His Majesty's pleasure shall from time to time direct.
>
> (*King's Regulations and Admiralty Instructions*, Chapter 1 Article 5)

Much has been written of the warships of the Royal Navy, social histories, campaigns and general histories, but information on naval uniforms of any period has always been somewhat sparse and lacking in detail, particularly with regard to ratings' clothing. This work sets out to fill the gap in current writing by providing a summary of the clothing and personal equipment used by the Officers and Ratings of the Royal Navy during World War II and the decade preceding that conflict. Whilst duly acknowledging the contribution of the Women's Royal Naval Service, the WRNS is not covered in this work, but is explored in detail in this author's title *World War II British Women's Uniforms.*

Whilst every effort has been made to standardize the terminology used in this work, it should be understood that contemporary RN publications can often use a number of designations to describe the same item. An extreme example of this confusion is the high neck white woollen sweater issued to submariners. It is variously designated as the Frock, White; Sweater, White; Submarine Sweater; Submarine Frock; and Frock, Woollen, Submarine Service ('Frock' is the archaic term for 'A woollen jersey worn by sailors'). Other than for isolated examples where more than one designation is used in official documents, the alternative names are listed in the text.

On occasion I have felt that detailing changes made outside the period covered by this work would enhance the overall understanding of a subject, or dispel some well entrenched myths and misunderstandings. It is my hope that anyone reading this work will learn something from my efforts, and above all will acquire a better understanding of the uniforms and effects of the British sailor from 1930 to 1945.

A group of visual signallers in No.1 uniform with Sennit hats pose for a photograph during a Sunday in harbour. *c.*1900.

PART ONE
The Historical Development of Royal Navy Uniforms 1856 to 1930

The first uniform regulations for the Royal Navy were issued by Lord Anson in 1748 to overcome 'the inconvenience arising from the want of an establishment of rank and precedence between His Majesty's Sea and Land Officers, as well as from the want of a due distinction among the Sea Officers themselves'. At the time Britain was the only European nation that did not have a regulation uniform for naval officers. The new uniform regulations provided for an elaborate and ornate Dress uniform for formal wear, and an Undress uniform for everyday wear.

In 1767 the Undress uniform was adopted as the Dress uniform, and a simpler Undress uniform was introduced because the cost of the gold lace used in the old Dress uniform had become prohibitive. In 1783 a highly elaborate Dress uniform with slash cuffs was provided for Flag Officers, but this was simplified only four years later again due to the expense. The cost of Britain's countless wars brought about changes to the uniform in 1795 when a plain blue frock coat was introduced for Undress wear, and epaulettes were introduced for senior naval officers. Dress uniform facings were also changed to blue. The use of epaulettes was extended to junior officers when in 1812 a single epaulette was authorized to be worn on the right shoulder by lieutenants. Dress uniform facings reverted to white in the same year and remained so, apart from the period between 1830 and 1843 when they were red.

Uniform regulations in book form and containing illustrative plates were first published in 1825; this edition introduced a blue cloth cap and Round Jacket, without skirts and worn open, as Undress uniform for wear at sea. Dress uniform Cocked Hats were henceforth to be worn uniformly in the 'fore and aft' at all times. In 1828 elaborate lapels disappeared from coats, which were in future to be worn buttoned up, and pantaloons or trousers replaced the white breeches previously worn. In 1856 white trousers were discontinued for home service use, and a cap badge on mohair band was introduced for wear by Officers on the cloth cap.

In 1861 the rank of Sub Lieutenant replaced that of Mate, necessitating a revision of commissioned rank structure and the addition of a rank stripe to that worn by Lieutenants, Commanders and Captains, who henceforth were to wear two, three and four stripes respectively. In 1863 distinction colours were introduced for wear between the rank stripes of civil branch officers. Also in 1863 a white cap cover was introduced for hot-weather wear on black caps.

In 1877 senior lieutenants of eight years' seniority were granted a half stripe to be worn between the two lieutenants' stripes, although the rank of Lieutenant Commander was not established until 1914.

The year 1880 saw the introduction of a new Ship's Jacket, to be worn at sea during bad weather or during the hours of darkness. It was only in use for five years. Between 1885 and 1889 a blue cloth version of the tropical tunic was worn, and in 1889 the Monkey Jacket made its first appearance. As the Undress coat it was still in use during World War II.

Illustrated uniform regulations in book form were published in 1825, 1879, 1891, 1924 and 1937.

Uniform regulations for seamen were somewhat slow in their formulation. By the end of the eighteenth century a 'best' uniform was being worn by many seamen, namely long trousers, short blue jacket with white taped seams, neckerchief and a round tarred straw or canvas hat. For working dress, canvas trousers and a variety of shirts and headwear were worn. Hat ribbons with the name of the ship painted upon them were in limited use by the start of the nineteenth century. Ultimately, the dress of the ordinary seamen was for quite some time the choice of the individual or the commanding officer of the ship.

However, by the 1850s a degree of uniformity had been established for seamen: a blue collar was being worn, and a round cap with named ribbon gave the seaman an immediately recognizable appearance. In 1857 the first 'Uniform Regulations for Petty Officers, Seamen and Boys' were established. Authorized uniform consisted of a Pea Jacket, serge and duck frocks and trousers, Sennit Hat, cap, a silk handkerchief and a comforter. The design of the regulation jackets and other clothing closely followed the clothing types already in widespread use. The white duck clothing was worn with the Sennit Hat as hot weather rig, while the Sennit was also worn with the blue serge frock but covered with black-painted 'Holland' fabric. In 1880 the seaman's jumper was introduced, in white duck and blue serge; the jumper was to be worn outside the trousers, unlike the frock that was to be worn tucked inside the trousers.

A basic selection of badges was worn on the regulation uniform: white badges on blue clothing, and blue badges on white clothing. The white badges were changed to red in 1860. Badges consisted of a crown and anchor encircled by a laurel wreath for Chief Petty Officers (CPOs), while 1st Class Petty Officers (POs) wore a crown and crossed anchors, 2nd Class Petty Officers wore an anchor surmounted by a crown, leading seamen wore an anchor, gunnery instructors wore a gun with crossed cutlass and rifle, seamen gunners 1st Class wore a gun and crown, and seamen gunners 2nd Class wore a gun badge. Other badges were produced over the coming years: in 1880 shipwrights wore an anchor, and in 1884 torpedoes were added to the gunnery specialization and subsequently new badges were introduced with torpedoes and guns in combination with stars and crowns.

In 1873 Chief Petty Officers were granted the use of gilt

buttons to replace the black horn buttons previously used. 1890 saw the introduction of different classes of uniform for ratings Classes I, II and III: Class I uniform was the 'fore and aft' rig of jacket with gilt buttons and peaked cap worn by CPOs. Class II uniform was the 'square rig' of jumper and bell-bottom trousers worn by seamen. Class III uniform was a 'fore-and-aft' rig of jacket with black horn buttons and peaked cap worn by men not dressed as seamen.

In 1920 all POs with four years' seniority were granted the wearing of Class I CPOs' style of uniform with gilt fastening buttons and two gilt cuff buttons. This was a massive boost for the prestige of seamen POs who had previously worn Class II square rig. The CPOs' cap badge was henceforth to be worn by POs, and a new badge with the addition of a laurel wreath was introduced for CPOs. Henceforth CPO non-substantive badges were to be worn on the lapels, thus distinguishing them from POs who wore their badges on the right sleeve. In 1925 the period of four years' acting was dropped to one year before confirmation as PO. The use of gilt cuff buttons by CPOs was extended to all branches under AFO 2713/25. It stated that the wearing of three large gilt buttons on the cuffs of jackets and White Tunics, previously restricted to certain CPOs only, was to be extended to all CPOs. The changes were to be made at the individual's expense.

Dating from the 1870s, this photo shows a sailor wearing the Seaman's Round Jacket over the White Frock with the White (Sennit) Hat.

*c.*1880: the Seaman's Cap worn with the Blue Serge Jumper.

Student Engineering Officer, morning dress 1880s.

Warrant Officer with sword belt, morning dress 1880s.

Warrant Officers, 1890s. Of note are the cuff buttons and the breast pockets on the Monkey Jacket.

RNR cuff buttons for a Warrant Officer.

A Victorian Full Dress Coat for an executive branch Lieutenant of the Royal Navy. The coat remained unchanged until its demise as an item of dress at the outbreak of World War II.

A Lieutenant RN wearing Full Dress uniform during the 1920s.

Victorian Full Dress uniform cuff decoration for a Commander of the RNR, civil branch.

Victorian Full Dress uniform epaulette for Sub Lieutenant RNR civil branch (silver piping).

A young RN officer wearing the open-collared White Coat, an optional item of tropical clothing worn instead of the White Jacket with stand collar.

A fine study of a Victorian Petty Officer armourer wearing No. 1 Dress.

A Petty Officer gunner from HMS *Prince George*, a Majestic class battleship (1895–1921). His collar appears to have a single broad edging, but in reality the three tapes are stitched so close as to appear as one. He carries the bosun's call in the knife pocket.

A volunteer seaman of the RNR *c*.1900. The Reservist wears the brown leather Sea Service Equipment and is armed with a Lee-Metford rifle. The collar tapes have been given a waved edge rather than a straight one.

*c.*1910. Stokers wearing the White Duck uniform with Sennit Hat as No. 6 Dress for a Sunday in harbour. The sleeve of the duck uniform jumper was worn short with the cuff high above the wrist.

Seamen wearing the serge jumper with White Duck Trousers and Sennit Hat as the pre-1933 No. 7 or No. 8 Dress.

A Sub Lieutenant of the Royal Naval Division wearing Khaki Service Dress with both RN cuff rank and Army shoulder rank consisting of two pips, equating to Lieutenant.

ABOVE: RNR Reservist of the Royal Naval Reserve Trawler Section. His cap ribbon bears the lettering 'H.M. MINE SWEEPER', *c.*1916.

RIGHT: Royal Naval Division rating of the Great War wearing Khaki Service Dress, khaki seaman's cap and 1908 web equipment.

Ratings in No. 5 working dress drill with Canadian Ross rifles on board HMS *Royal Sovereign* during World War I.

At left is a winter cap with wire grommet, low crown and blue cloth top, and at right a summer cap with cane grommet, duck top and low crown. Alongside the popular commercial caps these patterns were in use from pre-World War I through to the early 1930s.

A rating wearing the blue cap with H.M.T.B.D. (Torpedo Boat Destroyer) ribbon.

Mine Clearance Service arm badge worn on the lower left cuff. The badge was worn by members of the RN service formed in February 1919 to clear sea mines. A Mine Clearance Service cap ribbon was also worn by Class II Ratings.

A Great War period Class III Rating wearing the long coat with the upper buttons worn exposed at the lapel.

A Rating wearing the wool overcoat and blue cap with 'HM MINESWEEPER' ribbon.

Interior of the early summer cap with cane grommet, showing the plain crown with the owner's name stamp and fabric chinstay.

PART TWO
Royal Navy Uniforms 1930 to 1945

It was a requirement of King's Regulations Article 165 that uniform be worn by all RN, RM and WRNS personnel when inside naval establishments and when on short leave. When on leave for more than forty-eight hours' duration it was permissible to wear plain clothes, except when proceeding to or from HM ships or establishments and the leave destination. It was not permissible to wear uniform at any time when entering or travelling to or from Eire, for which destination all personnel were required to wear plain clothes. This order was later extended to include Ratings proceeding on leave to Northern Ireland, and also to prohibit the carriage of uniform in baggage while on leave to these destinations, it being required that all uniform items were left in the ship or the establishment in which the rating was serving.

During World War II it was a requirement that respirators were to be carried at all times, and those men who had been provided with a personal issue of a steel helmet were similarly to carry it at all times.

1 Officers' Uniforms

This section provides the classification and orders of dress of Royal Navy officers' uniforms, and the occasions on which they are worn. The following numbered classes of Officers' uniform relate to those being worn in June 1939, a time when Britain was at peace, but preparing for war. The uniforms reflect the nature of the Royal Navy of the era, a Navy whose role had been flying the flag in all the corners of Empire and further afield, with all the pomp and ceremony required of such. Officers' Uniform Dresses were renumbered in 1934, and all dress numbers referred to in the text relate to the period 1939–45 unless otherwise detailed.

Numbered Classes of Officers' Uniform

Dress No. 1 Full Dress

The No. 1 Dress was used for state occasions, when receiving the King, the Queen or other crowned heads, and at other ceremonies or events where the senior Officer considered that the occasion warranted it. Full Dress uniform consisted of the Full Dress coat with epaulettes, laced trousers, Cocked Hat, sword, Full Dress sword belt and white gloves. Full regalia of orders, decorations, medals and stars of orders were to be worn. The Full Dress was obligatory wear for all Officers at state balls. Officers below the rank of Commander were authorized the use of the Frock Coat with epaulettes for other Full Dress events if they did not possess the Full Dress coat. If the Great Coat was required to be worn over the Full Dress or Frock Coats because of poor weather the epaulettes were not to be fitted to the Full Dress or Frock Coats.

Dress No. 2 Ball Dress

The No. 2 Dress was Ball Dress, and was used at official or public balls, dinners and evening receptions. It consisted of the Mess Jacket with epaulettes, laced trousers, white waistcoat and white gloves. The full regalia of orders, miniature badges, decorations, medals and stars of orders was to be worn. The use of the Cocked Hat, sword and Full Dress sword belt could be ordered when meeting foreign naval Officers. Officers of Flag Rank, Commodores and Captains could, at their discretion, wear the Undress tail coat in lieu of the Mess Jacket. Ball dress was worn by all commissioned Officers, subordinate Officers above the rank of Midshipman, commissioned Officers from warrant rank, and Warrant Officers.

Dress No. 2A Ball Dress Without Epaulettes

No. 2A Ball Dress Without Epaulettes was used at official or public balls, and at dinners and evening receptions of a less formal nature than those that required No. 2 Ball Dress. The uniform consisted of that used for No. 2 Ball Dress, but the Mess Jacket was worn without the epaulettes. The 2A Dress was introduced in November 1937.

Dress No. 3 Frock Coat with Epaulettes Dress

No. 3 Dress was used for formal occasions including the receiving of heirs to the throne or other members of English or foreign royalty *when masthead flags were hoisted*, courts martial, funerals, boarding foreign ships of war, when receiving visits from or visiting senior foreign Officers or officials, and ceremonial or other duties where a degree of formality was required. No. 3 Frock Coat with Epaulettes Dress consisted of the Frock Coat, epaulettes, Morning Waistcoat, blue or white trousers, Cocked Hat or Sun Helmet, sword, Full Dress sword belt and white gloves. The Undress sword belt was worn by Warrant and Commissioned Warrant Officers. Full regalia of orders, decorations, medals and stars of orders was to be worn. The Frock Coat was worn by all commissioned Officers, subordinate Officers above the rank of Midshipman, commissioned Officers from warrant rank, and Warrant Officers. No. 3 dress was not worn by Midshipmen and Cadets, who instead wore the Round Jacket.

Dress No. 4 Frock Coat Dress

No. 4 Dress was used for less formal occasions including the receiving of heirs to the throne or other members of English or foreign royalty *when masthead flags were not hoisted*, also Sunday divisions and Sunday in harbour after divisions, inspections by the Commander-in-Chief or senior Officer, Officer of the guard, ordinary occasions of duty or ceremony ashore, visits to the Commander-in-Chief or senior Officer, and when receiving visits from, or visiting senior foreign Officers or officials. On occasion the Frock Coat Dress was worn without the sword and belt; this included while on leave during daytime, and at dances and entertainment afloat or ashore during daytime. The Frock Coat Dress with Undress sword belt but without sword was worn by the Officer of the watch while in harbour. The No. 4 Dress consisted of the Frock Coat, Morning Waistcoat, blue or white trousers, cap, sword, Undress sword belt and white gloves. Regalia consisted of the ribbons of orders, decorations and medals. The wearing of stars of orders was optional. Epaulettes were not worn with No. 4 Dress. No. 4 Dress was not worn by Midshipmen and Cadets, who instead wore the Round Jacket.

Dress No. 5 Undress

The Undress uniform was the normal working attire of Officers. It consisted of the Undress Coat (popularly known as the Monkey Jacket or Reefer), blue or white trousers, cap, sword, Undress sword belt, and the ribbons of orders, decorations and medals. In colder weather the Morning Waistcoat was also worn. Prescribed use, with or without sword and belt, was for drills, exercises and occasions of duty afloat other than those for which another dress order was prescribed and for patrol, dockyard duties and landing parties. Without the sword and sword belt No. 5 Undress was worn by Officers at home ports going to or from their place of residence, and for all other ordinary day-to-day occasions ashore or afloat or in HM dockyards. The Undress No. 5 uniform was worn by all Officers. Midshipmen and Cadets did not wear the sword belt.

Dress No. 6 Mess Dress

For occasions when a formal dress was required but the Ball Dress was too ornate, the No. 6 Mess Dress was worn. It consisted of the Blue Mess Jacket (or for Flag rank Officers, Commodores and Captains the optional wearing of the Undress tail coat), White Evening Waistcoat, laced trousers, miniature orders, decorations and medals. No. 6 Mess Dress was worn for a variety of events, including dinner in harbour when dining with Flag Officers, Commodores and Officers of corresponding rank when Ball Dress was not worn. It was also worn for such occasions as evening dances and entertainment ashore or afloat, dinners given by Military or Air Force Officers when they were wearing their Mess Dress, and dinner at the tables of Captains or Officers in Command, at the messes of Royal Naval colleges, barracks, medical establishments, wardroom and gun-room messes when Officers of Flag or corresponding Military and Air Force rank wearing uniform were guests. Mess Dress was not worn by Midshipmen and Cadets, who instead wore the Round Jacket.

Dress No. 7 Mess Undress

The less formal No. 7 Dress consisted of the Blue Mess Jacket (or for Flag rank Officers, Commodores and Captains, the optional wearing of the Undress tail coat), Blue Evening Waistcoat, plain blue trousers, and the ribbons only of miniature orders, decorations and medals. No. 7 Mess Undress was worn for dinner at sea when at the tables of Flag Officers and Commodores, and in harbour when guests were not entertained. It was also used for dinner at the tables of Captains or Officers in Command, at the messes of Royal Naval colleges, barracks, medical establishments, wardroom and gun room messes. Mess Undress was not worn by Midshipmen and Cadets who instead wore the Round Jacket. No. 7 Mess Undress was removed from the initial outfit requirements of all temporary Officers and Warrant Officers in January 1942. The Undress Coat was to be worn by these temporary wartime Officers as Mess Undress.

Dress No. 8 White Full Dress

No. 8 Dress was used for state occasions, when receiving the King, the Queen or other Crowned Heads unless specially ordered otherwise, receiving of heirs to the throne or other members of English or foreign royalty *when masthead flags were hoisted*, or at other ceremonies or entertainments when the senior Officer present considered it desirable to do special honour to the occasion. The tropical No. 8 White Full Dress uniform consisted of the White Tunic with shoulder straps, white trousers, Sun Helmet, sword with Full Dress sword belt, white gloves, and orders, decorations, medals and stars of orders.

Dress No. 9 White Dress

White Dress was a less formal version of the White Full Dress uniform, and was to be worn for courts martial, funerals, boarding foreign ships of war, exchanging visits of ceremony with governors general and other senior foreign and functionaries, and for other occasions of duty and ceremony when White Undress was not sufficient. The White Dress uniform No. 9 consisted of the White Tunic with shoulder straps, white trousers, white cap or cap with white cover (or at the senior Officer's discretion, the Sun Helmet could be worn), the sword with Full Dress sword belt, white gloves and orders, decorations, medals and stars of orders. Interestingly if the Sun Helmet were authorized for wear there was no difference between No. 8 White Full Dress and No. 9 White Dress uniforms. Prior to 1934 White Dress had been numbered 8A.

Dress No. 10 White Undress

White Undress was the informal version of the White Dress uniform, and was to be worn for the receiving of heirs to the throne or other members of English or foreign royalty *when masthead flags were not hoisted*; also divisions on Sunday, inspections by the Commander-in-Chief or senior officers, visits to the Commander-in-Chief or senior officer, attending hospitals, Officer of the Guard (except when boarding foreign ships of war), ordinary occasions of duty, and ceremony ashore. Without the sword it was worn on leave in the daytime, and for dances and entertainment afloat or ashore in the daytime. On Sunday in harbour after divisions (with sword), and with the Undress sword belt, but without the sword, it was worn by the Officer of the watch in harbour. Stars of orders were optional for all of the preceding occasions. Stars of orders were not to be worn for the following occasions on which No. 10 Dress was used: drills, exercises and occasions of duty afloat other than those for which another dress is prescribed (with or without the sword, dependent upon the nature of the exercise or drills).

The No. 10 Dress uniform was also used for patrol, dockyard duties and landing parties and, again without the sword, all other ordinary occasions ashore or afloat or in HM dockyards. The No. 10 White Undress uniform consisted of the White Tunic with shoulder straps, white trousers, white cap or cap with white cover (or at the senior Officers' discretion, the Sun Helmet); also the sword with Undress sword belt, white gloves, and the ribbons of orders, decorations and medals. Stars of orders were optional wear for some of the more formal occasions on which the No. 10 uniform was to be worn. Prior to 1934 White Undress had been numbered 8B.

Dress No. 11 White Mess Dress

No. 11 Dress was worn for formal mess functions in tropical regions. It consisted of the White Mess Jacket with shoulder straps, White Evening Waistcoat, laced trousers and miniature orders, decorations and medals. No. 11 White Mess Dress was worn for dinner in harbour when dining with Flag Officers, Commodores and Officers of corresponding rank if specially ordered by the senior Officer, evening dances and entertainment ashore or afloat if specially ordered by the senior Officer, and dinners at the military or Air Force mess, and entertainments given by military or Air Force Officers when they were wearing their white Mess Dress. Prior to 1934 White Mess Dress was numbered No. 9.

Dress No. 12 White Mess Undress

White Mess Undress was worn for less formal mess functions in tropical regions. It consisted of the white Mess Jacket with shoulder straps, Blue Evening Waistcoat or kamarband, plain blue trousers, and the ribbons of miniature orders, decorations and medals. No. 12 White Mess Undress was worn in tropical regions for dinner at sea and in harbour unless No. 9 Dress was ordered by the senior Officer, at the tables of Flag Officers and Commodores. It was also used for dinner at the tables of Captains or Officers in command, at the messes of Royal Naval colleges, barracks and medical establishments, wardroom and gun-room messes. Prior to 1934 White Mess Undress was numbered No. 10.

Dress No. 13 Tropical Dress

Tropical Dress was worn on tropical stations for all non-ceremonial occasions at the discretion of the Senior Naval Officer. Tropical Dress consisted of white shirt with shoulder straps, white shorts, and white stockings and white buckskin shoes or blue stockings and black shoes or laced boots. It was worn with the cap with white cover, white cap or Sun Helmet as ordered by the Senior Naval Officer. No. 13 Dress was instituted in 1938. Officers appointed to the Mediterranean station were authorized to wear No. 13 Dress in May 1939.

Orders of Dress Introduced during World War II

Dress No. 5A Working Dress

The 5A Working Dress could be worn when working at sea and in aircraft at all times, and when on passage by air; also in ships in harbour, naval air stations and naval air sections during working hours and non-ceremonial occasions, but not by duty Officers. 5A Working Dress could also be worn by armed landing parties or during active operations if khaki was not worn; or when doing arduous or dirty work when other clothing might be spoilt. Notwithstanding these regulations, the use of No. 5A Dress was further governed by Commanders-in-Chief and Flag Officers Commanding who had discretionary powers to vary the rules for the wearing of No. 5A Dress in exceptional circumstances.

Dress No. 5B Working Dress for Air Crews

The 5B Working Dress for Air Crews could be worn at sea and in aircraft at all times. Also in ships in harbour, naval air stations and naval air sections during working hours, and non-ceremonial occasions, but not by duty Officers. As with No. 5A Dress, Commanders-in-Chief and Flag Officers Commanding had discretionary powers to vary the rules for the wearing of No. 5B Dress in exceptional circumstances. Working dress for air crews was not to be worn on shore outside naval establishments except when travelling by rail or road directly to or from a flight or duty connected with aircraft; nor was it to be worn during leave, or when proceeding to or from places of residence.

Dress No. 14 Action Dress

The Dress No. 14 Action Dress was to be worn at action stations, or ashore or afloat, on duty or leave, when ordered by the Senior Officer.

Studio portrait of an RN Lieutenant wearing No. 5 Undress uniform.

An RNVR Sub Lieutenant wearing the White Undress uniform Dress No. 10.

Naval Cadets 'stand easy' while undergoing sea training. The Cadets all wear dark grey flannel trousers. Grey flannels were unique to Cadets undertaking the three senior terms of their training.

The Army pattern Service Dress Jacket worn with the Khaki uniform. This uniform was worn during World War I by Officers serving ashore with the RND, RNAS and other specialist units. Although still listed in regulations, it was rarely used during World War II. BR81 1937

Uniform Landed During the War

During the Great War only Undress, Mess Dress, White Undress and White Mess Undress were worn. As it had done in 1914, the outbreak of war in September 1939 again brought about an immediate change in uniform requirements for Officers, as those serving at sea during a conflict had no operational requirement for the regalia of peacetime. The Admiralty also wished to reduce the heavy liability for compensation of expensive uniform articles when ships were sunk or clothing was lost due to enemy action, and had declared such in documents drawn up in June 1939.

AFO 2605 of 9 September 1939 ordered that a number of unnecessary uniform items were to be landed. Items no longer required were the No. 1 Full Dress, No. 2 Ball Dress, No. 2A Ball Dress without epaulettes, No. 3 Frock Coat with epaulettes, No. 4 Frock Coat Dress, No. 6 Mess Dress, No. 8 White Full Dress, No. 9 White Dress and No. 11 White Mess Dress. Only the No. 5 Undress, No. 7 Mess Undress, No. 10 White Undress, No. 12 White Mess Undress and No. 13 Tropical Dress were retained as a wartime uniform requirement. In January 1942 the No. 7 Mess Undress was removed from the initial outfit requirements of all temporary Officers and Warrant Officers, and the requirement for a walking stick was removed from all initial uniform outfits. These uniforms remained in abeyance throughout the war, with some orders of Dress never reintroduced.

AFO 2605 was updated in December 1939. The Boat Cloak was added to the items that should be landed '*at the first port*', but Officers were advised that the Undress Tail Coat, Watch Coat, sword and Undress sword belt could be retained on board at the individual Officer's discretion.

King's Regulations article 1697 allowed compensation for Officers and Ratings who had lost kit as a result of enemy action. However, during the war no claims were to be entertained for items lost at sea that should have been landed.

RNR and RNVR Uniform Requirements

BR 62, Regulations for the Royal Naval Reserve (Officers) 1937, provided details of the uniform requirement for reserve Officers. It was to be of the same pattern worn by Royal Navy Officers but with RNR or RNVR pattern cuff lace or turnbacks.

All Officers except Midshipmen, Chief Skippers, Skippers, Commissioned and Warrant Engineers were required to provide themselves with the following uniform items: No. 5 Undress, No. 6 Mess Undress and No. 4 Frock Coat dress. There was a further requirement for white drill or duck trousers, white cap covers and white shoes for Officers undergoing training between 1 May and 1 October. Officers appointed to the Home Fleet spring cruise required the above items plus White Tunics. Officers undergoing the special nine-month training period were appointed to ships of the Mediterranean Fleet on conclusion of their courses. From May to October full white uniform was required. Any items of RN Officers' dress appropriate to rank could be purchased by RNR and RNVR Officers.

RNR Officers attending the gunnery or torpedo qualifying courses were required to provide themselves with two pairs of white flannel trousers, a white jersey, a white muffler and black lace-up boots. When worn, these items were known as 'training rig'.

Chief Skippers and Skippers were required to provide themselves with the following items, of the patterns worn by Commissioned Warrant Officers and Warrant Officers RN: one Undress Coat, one waistcoat, one pair of plain blue trousers, one cap, one foul-weather hat, one pair of black boots, one long oilskin coat, one necktie, white shirts and collars. Upon mobilization or for service with the Fleet, Chief Skippers and Skippers were required to furnish themselves with three white tunics, three pairs of white trousers, one set of tunic buttons, one Sun Helmet, one pair of shoulder straps and one pair of white shoes.

Commissioned Engineers and Warrant Engineers were required to provide themselves with the following items when undergoing training on HM ships: one Undress Coat, waistcoat and trousers of best blue cloth, one Undress Coat, waistcoat and trousers of blue serge, two blue caps, two black ties, one pair of boots, and white shirts, collars and cap covers. A Frock Coat and sword belt were optional items.

RNR probationary Midshipmen appointed for four months' training with the Fleet were required to provide themselves with the following uniform: one Undress Coat cloth, one Undress Coat serge, one pair uniform trousers cloth, one pair uniform trousers serge, two pairs uniform boots, one pair of rubber seaboots, six white shirts, four evening shirts, twelve collars, one pair brown gloves, two black ties, one evening tie, one oilskin coat, one sou'wester, two uniform caps, four white cap covers, one pair evening shoes, one pair gaiters, one Round Jacket (optional), one Round Jacket waistcoat (optional), and one pair of trousers for wear with the Round Jacket. RNR Midshipmen, when appointed for short periods or when under training, were not required to wear a sword or dirk, but were allowed to wear a sword.

Classes of RNR Officers on the active list were as follows:

Executive Officers: Commodore, Captain, Commander, Lieutenant Commander, Lieutenant, Sub Lieutenant, Acting Sub Lieutenant, Midshipman, Skipper Lieutenant, Chief Skipper, Skipper.

Engineer Officers: Engineer Commander, Engineer Lieutenant Commander, Engineer Lieutenant, Commissioned Engineer, Warrant Engineer.

Accountant Officers: Paymaster Commander, Paymaster Lieutenant Commander, Paymaster Lieutenant, Paymaster Sub Lieutenant.

Officers' Uniform Items in Detail

Full Dress Coat

The Full Dress Coat was worn by all commissioned Officers, subordinate Officers above the rank of Midshipman, commissioned Officers from warrant rank, and Warrant Officers. It was a blue cloth double-breasted coat with two rows of eight buttons. Highly ornate, it had a white mandarin-type collar trimmed with gold oak leaf or lace according to rank. The cuffs were adorned with a white 'slash', having three regulation buttons embroidered with gold oak leaves for flag Officers, or a gold lace border for other Officers. The slash was placed over the corresponding cuff rank lace. The shoulders had fittings for the use of epaulettes. Warrant Officer's Full Dress Coats were not worn with epaulettes. The Full Dress Coat was not worn by Midshipmen and Cadets, who instead wore the Round Jacket. BR81 1937 stated:

> Commanders of more than four years' seniority on 1 July 1930, and Officers below the rank of Commander, need not provide the Full Dress Coat.

The Full Dress Coat was worn with No. 1 Dress.

Round Jacket

The Round Jacket, worn by Midshipmen and Cadets, was a short, single-breasted blue cloth coat with a mandarin

An RNVR Lieutenant wearing No. 5 Dress on the bridge of a Harbour Defence Motor Launch. The rank lace is the half-cuff war economy pattern introduced in August 1941.

At left is the Full Dress Coat as worn by a Flag Officer or Commodore First Class. *Centre* is the coat worn by other commissioned Officers. *At right* is the coat worn by Commissioned Officers from Warrant Rank or Warrant Officer. BR81 1937

Detail of the cuff of the Round Jacket worn by Officer Cadets showing the three buttons and notched holes of black twist.

An Officer Cadet wearing the Round Jacket with Morning Waistcoat. The Round Jacket was worn by Midshipmen and Cadets on occasions for which other Officers would wear the Full Dress Coat, Frock Coat or Mess Jacket.

Young Officer Cadets under instruction at a naval college. The Cadets wear the Undress Coat with white trousers. Cadets wore a lanyard with No. 5 Dress with the free end tucked into the breast pocket.

collar and seven buttons, worn unfastened. At each cuff there were three notched holes of vertical black twist with buttons. It was worn over the morning waistcoat. The Cadet's notched hole with twist and button, or the Midshipman's turnback and button, was worn at the collar. The Round Jacket was worn on occasions for which other Officers would wear the Full Dress Coat, Frock Coat or Mess Jacket. Prior to September 1937 the collar of the jacket had been closed with a hook and eye fastening. AFO 1980/37 removed the requirement for the fastener, with all new production being made without the hook and eye.

In April 1936 AFO 909 announced that Paymaster Cadets and Midshipmen who entered the service after 1 September were in future to wear the Round Jacket on occasions when it was worn by other Cadets and Midshipmen. Paymaster Cadets and Midshipmen had previously worn the Frock Coat.

The Round Jacket was worn with No. 1, No. 2, No. 2A, No. 3, No. 4, No. 6, No. 7, No. 11 and No. 12 Dresses.

Frock Coat

The Frock Coat was worn by all commissioned Officers, subordinate Officers above the rank of Midshipman, commissioned Officers from warrant rank and Warrant Officers. It was a blue cloth double-breasted coat with a padded turn-down collar. It had two rows of four buttons, with

A pre-1933 Frock Coat laced to a Lieutenant Commander Engineer RN. The double row of five buttons was used until 1933. At right is the rear of the coat.

Sub Lieutenant RNVR dressed in Officers' No. 3 uniform. The long Frock Coat is worn with Cocked Hat, epaulettes, white gloves, dress sword belt and sword.

Gold lace epaulettes with anchor and chain device as worn by Sub Lieutenants and Commissioned Officers from Warrant rank, with No. 1 Full Dress, No. 2 Ball Dress and No. 3 Frock Coat with epaulette dress.

three buttoning. Prior to July 1933 the coat had two rows of five buttons, four to button. Rank lace and coloured distinction cloth fill was worn at the cuff, according to rank and branch. The shoulders had fittings for the use of epaulettes, except for Warrant Officers, whose Frock Coats were worn without epaulettes. A hook to support the sword belt was fitted at the left waist. Prior to April 1936 the Frock Coat had also been worn by Paymaster Cadets and Midshipmen. Other Midshipmen and Cadets did not wear the Frock Coat but wore the Round Jacket instead. The Frock Coat was landed under AFO 2605/39.

The Frock Coat was worn with No. 3 and No. 4 Dress.

Sub Lieutenant RNVR (Sp) wearing the No. 4 Dress as off-duty daytime wear with white trousers, buckskin shoes, white gloves and white cap cover. The Officers' walking stick was usually carried with this dress, worn without the sword or sword belt.

Admiral Sir Charles Forbes inspects a Royal Marine guard on board HMS *Nelson* following his appointment as Commander-in-Chief Home Fleet in 1938. Admiral Forbes wears the No. 4 Frock Coat Dress with sword and Undress belt.

The No. 4 Frock Coat Dress, worn by the Captain of HMS *St Vincent c.*1937 without sword and belt, as daytime off-duty dress. He carries white gloves and walking cane.

Undress Coat

The Undress uniform was worn by all Officers as daily working dress. The double-breasted Undress Coat had a padded turn-down collar and two rows of four buttons. At each side of the lower skirt there was a flapless pocket, the top being in line with the lowest button. A small pocket was located on the upper left breast. At each side of the lower skirt side seams there was a 5in vent. Rank lace and coloured distinction cloth fill was worn at the cuff. Midshipmen wore the turnback and button on the collar, Cadets wore the notched hole with white twist and button. Paymaster Midshipmen and Cadets wore the white branch distinction colour on the cuff.

Despite the wartime austerity measures applied to both military and civilian clothing, within the RN it was Ratings' clothing that was mainly affected, other than the introduction in 1941 of half-width cuff rank for Officers. Despite the pressures on manufacturers in regard to both availability of fabric, production time and output of garments, the Admiralty insisted that hand finishing was essential in many stages of the production of Officers' uniforms, compatible with the high standards of tailoring to which naval Officers were accustomed. While departure from bespoke standards was allowed in less important details of construction, the Undress Coat required hand finishing or hand stitching of the shoulder padding, lapels, linings, collar, buttonholes and distinction lace. It also retained the superfluous second row of hand-stitched buttonholes on the inner turn, despite the fact that they were never used and took quite some effort to fabricate.

The Undress Coat was worn with No. 5 Dress.

RNR Midshipman's Undress Coat with blue collar 'turnback' and notched hole worn by Midshipmen. Turnbacks worn by RNVR were maroon, those worn by RNR were blue and those of RN midshipmen were white.

RNVR Midshipman showing the collar 'turnback' and notched hole worn by all Midshipmen.

Undress Coat laced for a Warrant Officer Paymaster.

HMS *Valiant* crash cutter winners 1934. The Officers wear No. 5 Dress with white trousers. The Acting Petty Officer wears No. 7 Dress, which in 1934 was worn with white drill trousers.

A Lieutenant (E), engineering officer of the RNVR, wearing No. 5 Dress uniform with brown gloves. The rank lace distinction inlay for engineers was purple.

No. 5 Undress uniform worn by a Lieutenant Commander RN, a veteran of the Great War.

The navigating officer of HMS *Rodney* working at the chart table. The Undress Coat – also known as a 'Reefer' or 'Monkey' Jacket – was normal working dress for Officers.

RIGHT: Detail of the medal ribbons worn by the Lieutenant Commander RN British War medal, Victory medal and the Albanian 'Commander of the Order of Skanderbeg'. This order was awarded to only a very few recipients, including Lieutenant Commander Conway Vivian Dodgson, who was granted royal licence by the King to wear the foreign award.

A No. 5 Undress Coat laced to a Commander of the executive branch RNVR 'Wavy Navy'.

Undress Coat laced to a Captain of the Royal Navy Reserve.

Undress Coat laced to a Commodore Second Class of the Royal Navy, the highest ranking line officer. The ribbons are the 1914–15 star, British War medal and Victory medal.

A Sub Lieutenant of the RNR shooting an object with a navigational sextant. He wears No. 5 Undress with white scarf around his neck.

Chaplains' Dress

Until 1931 RN chaplains wore normal ecclesiastical dress with no distinctive marks to identify them from their civil counterparts. In April 1931 a new black scarf was introduced for chaplains of the RN and RNVR under AFO 795, to be worn when officiating at naval, military or RAF services. The scarf, which was optional, was embroidered at each end with an anchor attached to a cross by its anchor chains and surmounted by a crown. Medals or ribbons could be worn on the left side of the scarf. The letters 'RNVR' were embroidered below the device for scarves worn by chaplains of the RNVR. When RN Officers were in white uniform chaplains were authorized to wear a plain White Tunic with black Maltese crosses fitted to each side of the collar.

Chaplains tended to wear an ordinary jacket at times when naval Officers were wearing Mess Undress. Regulations required that they wear a clerical court coat, and they were reminded of this under AFO 1105/39, which advised that the coat was to be worn with silk cassock, waistcoat and knee breeches at all times when Officers were wearing Full Dress, Ball Dress or Mess Dress.

In September 1939 moves were made to allow the wearing of naval uniform by chaplains. The requirement was brought about by the need to be easily recognized as a military officer, something that was not possible in plain clothes. It was also felt that while Navy chaplains were sharing in the danger of warfare with their fellow Officers, there was nothing to connect them with the service in the public mind, and the wearing of naval uniform would strengthen the bond between chaplains and all ranks and Ratings in time of war.

In 1940 a blue cloth cap and a chaplain's cap badge was authorized under AFO 2106. The cap was of the standard Officer's style but it had a cloth-covered peak with a ¾in mohair braid edging. The cap badge consisted of a laurel wreath embroidered in black silk edged in gold, surrounding a silver fouled anchor all embroidered on a blue cloth ground, surmounted by a crown of gold and silver embroidery. It was to be worn with a plain Reefer (Monkey) Jacket with black buttons and no distinguishing marks, plain trousers and a Great Coat with no distinguishing marks. In 1942 AFO 3480 advised that for the period of hostilities chaplains could wear a blue reefer jacket with Officers' gilt buttons and without rank. The reefer jacket was of the RN Officers' Undress Coat pattern.

In tropical uniform chaplains were to wear approved Officers' dress with chaplain's shoulder straps bearing the Maltese cross. In white uniform with clerical collar an open-neck white coat with naval buttons was to be worn with black shoulder straps bearing a gilt Maltese cross.

When required to wear Battledress, chaplains were to wear the clerical collar and stock and not a khaki shirt and tie. The Maltese cross was worn on the shoulder straps with Royal Navy shoulder flashes. A khaki field service cap with bronze naval Officer's badge was to be worn as head-dress. Royal Navy chaplains serving with Royal Marine units were authorized to use khaki service dress with bronze naval buttons and Royal Navy shoulder flashes. A khaki service-dress peaked cap with chaplain's cap badge on a black band, or the black cap with khaki cover, was approved for wear with this dress. KD was to have bronze buttons and badges.

A Royal Navy Chaplain wearing the blue Reefer Jacket with clerical collar and the chaplain's cap and badge authorized for wear in 1940; the authority to wear a naval Reefer Jacket was granted in 1942.

The special cap badge authorized under AFO 2106/40 for optional wear by all RN and RNVR chaplains with the chaplain's pattern cap.

The black Maltese Cross insignia worn on both collars of the white uniform jacket by chaplains. BR91 1937

RIGHT: The special device worn by chaplains of the RN on the black silk scarf. The scarf was introduced under AFO 795 of April 1931. An identical scarf but with the letters 'R.N.V.R.' embroidered below the device was worn by chaplains of the RNVR.

A Roman Catholic chaplain of the Royal Navy conducts Mass on a major warship, 1941.

Officers' Working Dress

AFO 1036/43 stated that the Admiralty was not prepared to authorize the adoption of blue BD (Battledress) for naval Officers, and that blue BD was not to be worn. However, the order was often ignored. Despite their lordships' previous insistence that BD would not be adopted by the Navy, AFO 4669/43 authorized the introduction of a new BD-style wartime working dress for Officers, the No. 5A Working Dress. Made on the lines of Army BD it was expressly stated that it was not Battledress, was not to be called such, and had to be referred to as 'Working Dress'.

Details of the Working Dress were first promulgated in AFO 4669/43, but stocks were not available from Officers' clothing depots until January 1944. The blouse had two flap top plain breast pockets with gilt buttons. Three gilt uniform buttons fastened the front with three plain buttons for fastening and adjustment at the waist. Working Dress trousers were similar to those worn with No. 5 Dress, with two side pockets and two hip pockets. Both the blouse and trousers were available from naval stores in eighteen sizes, but Officers were free to purchase made-to-measure Working Dress from commercial sources.

The original size range did not prove adequate for Officers of unusual height or build. Under the provisions of AFO 5217/44, Officers who were unable to equip themselves with suitably sized Working Dress could provide their measurements on a form S.88, and upon payment to their accountant Officer a made-to-measure suit could be ordered from stores. Complete suits could be ordered, or blouse and trousers individually, but it was advised that if blouses and trousers were ordered separately there was a risk that the fabric colour of the two garments might not match.

Detail of the Lieutenant Commander RN rank straps on the Working Dress blouse. The distinction lace is of brown silk rather than gold lace, of a pattern available for purchase from naval outfitters

The 5A Working Dress uniform worn by a Lieutenant Commander RN. The working dress uniform was introduced in 1943 as a practical alternative to the Undress uniform.

Following the introduction of the Working Dress, some difficulty was encountered in attempting to provide gold rank lace in the huge quantity required to equip blouses with shoulder straps. It was therefore announced that straps for use with the Working Dress could be fitted with lace made from woven thread, and that such straps would be made available from service stocks or naval outfitters. Until such straps were available Officers were authorized to use the straps from their white uniform. By May 1944 sufficient woven straps had been produced for them to be made available for sale. The straps were of matching fabric $5\frac{1}{4} \times 2\frac{1}{4}$in, and were available from naval stores in the ranks of Warrant Officer to Captain RN, RNR and RNVR Turnbacks for Midshipmen were also produced.

Branch distinctions were initially only Executive, Accountant, Engineer, Electrical and Air Branch. Supplies of straps bearing coloured distinction of other branches were not to be stocked, but could be specially supplied as required. However, by 1945 stock had been increased, and medical, dental, special branch, instructor, shipwright and ordnance were available from victualling yards.

The Officers' Working Dress uniform was worn as No. 5A Dress.

Admiral Sir Bertram Ramsay and Deputy Supreme Commander SHAEF, Air Chief Marshal Sir Arthur Tedder on the Normandy beach-head during June 1944. Ramsay wears a privately tailored version of the 5A Working Dress with pleated breast pockets.

5A Working Dress jacket, as worn by a submariner Lieutenant Commander RNR, shown with the convertible collar closed at the neck. It is fitted with blackened metal buttons.

The special woven lace rank straps introduced in May 1944 for use with the 5A Working Dress. The backing fabric matches that used for the 5A uniform. This strap is a pattern 1007R for a Lieutenant Commander RNR. The strap buttons have been blackened.

A Submarine Commander wearing the No. 5A Working Dress uniform. Submariners' approach to uniform was generally lax, but the No. 5A Dress was well suited to wear in the confines of a sub. No rank straps are worn on this blouse.

Typical fabric label fitted to Lottery & Co. garments, in this case the Working Dress trousers. Unlike Ratings' labelling that was always paper, officers' clothing provided from Slops initially used fabric labels.

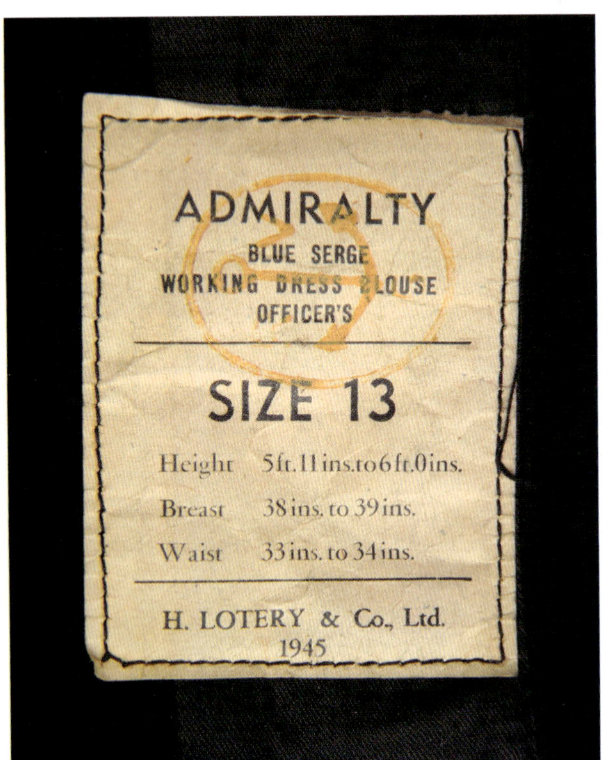

A 1945 dated Lottery & Co. paper label fitted to a Working Dress blouse. As an economy measure paper labels replaced fabric ones on Officers' clothing in 1945.

Air Branch and Fleet Air Arm

In October 1925 Royal Naval Officers serving with the RAF as members of the Fleet Air Arm were granted a badge for wear on the left sleeve. The badge was composed of a silver anchor and cable surrounded by a laurel wreath superimposed on the wings of an albatross in gold. In 1938, AFO 2885/38 announced the formation of an Air Branch of the Royal Naval Reserve, and on 24 May 1939 the Fleet Air Arm was returned to Admiralty control.

In March 1938 AFO 635 introduced a new badge to be worn by Officers of the Air Branch. It consisted of a capital A in silver within a gold-embroidered wreath. The badge was to be worn above the rank lace curl on the left sleeve, or in a corresponding position in the case of Midshipmen, and on shoulder straps when worn. This particular badge was short lived, as in April 1939 AFO 908/39 introduced new badges for Officers qualified as pilots in the Fleet Air Arm, and for Air Branch Officers of the RN and RNVR. The 2⅞in pilot's badge consisted of wings of gold embroidery, much broader than the preceding pattern, having in the centre a silver anchor and cable surrounded by a gold-embroidered wreath. The new wings were surmounted by a crown. The wings were to be sewn in position ¼in above the curl of the rank lace of the left sleeve, or in a corresponding position in the case of Midshipmen. On white and KD uniforms the wings were to be fitted, using a pin attachment, 1¼in above the left breast pocket, or immediately above medal ribbons if worn. The wings were not to be worn on tropical shirts, shoulder straps, Watch Coat, Great Coat or waterproofs. The new badge for Air Branch Officers consisted of a capital 'A' ½in in height to be worn inside the rank lace curl on both sleeves and on both shoulder straps when worn. It was to be worn in a corresponding position in the case of Officers who did not wear the curl.

In 1942 the Admiralty introduced a badge for observers. The badge consisted of gold wings having in the centre a fouled anchor surrounded by the letter O in silver rope surmounted by a crown. In May 1943 a new badge for Warrant Air Officers (Air Gunner) was approved. The badge resembled that of the observers but with a narrow gilt rope ring instead of the letter O.

The Fleet Air Arm pilots' wings of the pattern in use between 1925 and 1939. BR81 1937

Air Branch qualification badges. Above: Fleet Air Arm pilot wings of the pattern introduced in 1939; below: The observers' wings first introduced in 1942. Pin-backed versions were worn on the white uniform.

Warrant Air Officers (Air Gunner) wings as approved in May 1943. The badge resembled the observers' wings, but with a narrow gilt rope ring instead of the letter O.

Left sleeve cuff lace of a Sub Lieutenant of the RNVR Air Branch. Fleet Air Arm pilots' wings were worn above the lace curl on the left sleeve only of the Undress Coat.

Undress Coat laced to a Sub Lieutenant pilot of the RNVR Air Branch; the Undress uniform was the normal daily work dress and flying clothing of FAA officers.

Lieutenant pilot K. A. Remington wearing the No. 5 Dress with the Mk.I motorcycle despatch riders' steel helmet.

Working Dress for Air Crews

A new uniform for air crew was approved in March 1945. AFO 1283 gave details of the uniform to be worn by Officers and Ratings. The two-piece ensemble was similar to the Army's Battledress Serge but made in 'Navy serge, blue, No. 4'. The blouse was single breasted with a fly front that could be buttoned to the neck. The neck was tailored to be worn open with a tie, but it could be closed with a single hook fastener and a cloth tab. The blouse had two external breast pockets with plain front and flap top, and two internal breast pockets. An integral waistbelt and toothed buckle secured the blouse at the waist. The sleeve cuffs had fly-button closure and the shoulders were fitted with shoulder straps. The trousers were also similar to the Army's Battledress Serge with a thigh pocket, two side pockets and two hip pockets. At the waist, buttons provided fitting for braces, and buttons at the rear waist married with buttonholes on the rear of the blouse, securing the two together.

The uniform provided for Officers and Ratings was the same. However, the Officers' uniform was called No. 5B Dress and the Ratings' version was the 3A Dress. It was to be worn at sea and in aircraft at all times. It could also be worn in ships in harbour, in naval and dockyard establishments, and on naval air stations and naval air sections during working hours or on non-ceremonial occasions. It was not to be worn by duty Officers, and it was not normally to be worn ashore outside naval establishments, on leave or by Officers proceeding to or from their places of residence. The 3A Dress was to be added to the compulsory kit of all rating pilots, observers and telegraphist air gunners (TAGs), and a free issue was made. Officers would be able to purchase 5B Dress from service stocks or private outfitters. Prior to the introduction of the 5B Dress FAA Officers had unofficially adopted a blue serge BD style of uniform for use as flying clothing. Although against Admiralty regulations, the privately tailored BD provided a functional flying dress superior to the Undress Coat that was the authorized flying clothing.

Officers' flying badges were to be worn above the left breast pocket, and shoulder rank straps were to be secure-

A privately purchased flying uniform badged to an RNVR observer Lieutenant. The uniform is based on the Army's 1937 pattern Battledress. Despite Admiralty orders banning the use of Battledress, this style of uniform was commonly used as flying clothing within the Air Branch. The No. 5B Working Dress for Air Crews was based on this ensemble.

RNVR Air Branch rank badges worn on the observer Lieutenant's Battledress uniform. The rank lacing is khaki braid with the Air Branch 'A' embroidered on drill fabric, fitted to a wool slide.

War economy pattern rank straps for a Lieutenant (Air) for wear with the 5B Working Dress for Air Crew. These rank straps were to be stitched directly to the uniform shoulder straps.

ly stitched to the blouse shoulder straps. Ratings were to wear substantive and non-substantive badges in the usual manner. Medal ribbons were not to be worn with 5B or 3A Dress.

For details of the headwear to be worn with the No. 5 Dress, *see* the section 'Beret, Working Dress, Air Crews'.

The Officer's Working Dress was worn as Dress No. 5B.

The 5B Working Dress for Air Crew blouse, approved in March 1945. The 5B Dress was the Admiralty version of the Battledress that had been worn unofficially by aircrew officers since mid-war. PAUL LAIDLAW, WDMILITARIA

An interesting relaxed group of Officer and Rating pilots including RN, RNR and RNVR Sub Lieutenants and an Acting PO Rating pilot wearing his wings on his upper right sleeve. The flyer at extreme left wears the Sidcot suit.

Pilot qualification wings worn above the left breast pocket on a privately purchased Battledress-style flying blouse fitted with a fur collar from the Sidcot suit.

A Royal Navy Sub Lieutenant dressed in the No. 10 White Undress uniform with Sun Helmet.

Officers' White Uniform

Royal Navy Officers of the eighteenth and early nineteenth centuries had long complained that the blue wool uniform worn in British waters was also worn in tropical climates. A white cotton cap cover had been introduced for hot weather use in 1863 but it was a far from adequate concession to the Royal Navy's requirement for a dedicated tropical uniform.

In 1885 the Admiralty finally approved a tropical uniform that included a white cotton drill five-button single-breasted tunic with a stand- (mandarin-) style collar. Initially rank was worn on the cuff as white silk bands with branch distinction colours where appropriate; this was changed to shoulder straps in 1891.

In 1924 the Admiralty Board had refused to consider proposals then being presented for a more functional tropical working dress for Officers. However, by 1930 a new white tropical working rig had been approved by the board, but the dress was not accepted by the King, who had the final word on all such matters. Two years later the Commander-in-Chief East Indies proposed that a Khaki tropical uniform be adopted for RN Officers, but again the proposal did not meet the King's approval. The following year His Majesty did, however, approve the use of khaki shirt and shorts by Officers serving on submarines on the China Station, and the use of white shorts and stockings by Officers when Ratings were dressed in tropical rig. In 1935 the use of this rig was extended for wear on the America Station.

In the same year Officers on the East Indies Station adopted a 'No. 10A' Dress that consisted of a white open-neck cricket shirt with short sleeves and shoulder straps, white shorts, white shoes with white stockings, and blue garter tab for wear at sea and at Trincomalee, or with black shoes and blue stockings at Diyatalawa. Regulation No. 10 Dress was to be worn on Sunday mornings. In January 1937 the East Indies '10A' Dress was adopted on the China Station. A month later it was officially proposed that the local '10A' Dress used on the East Indies Station should be adopted as Officers' tropical rig throughout the service, but without the blue garter tab, for wear at sea and when no duties requiring a more formal dress were being undertaken. The proposal was accepted in principle, and the new dress was gazetted on 21 January 1938 with an amendment to BR 81 *Uniform Regulations* being published the following month. The amendment listed the new dress as the No. 13 Tropical Dress.

The introduction of the new tropical working rig in February 1938 saw Officers wearing a comfortable open-neck shirt with white cotton shorts, blue or white stockings and black or white shoes or lace-up boots. The uniform was to be worn with the cap and white cover or the Sun Helmet. But it was not until AFO 1290 of May 1939 that Officers appointed to the Mediterranean Station were authorized the wear of No. 13 Dress.

No. 8 White Full Dress uniform worn by a Royal Navy Lieutenant on the occasion of his wedding held in Hong Kong during the inter-war period.

HM King George VI wears the White Undress uniform on board the cruiser HMS *Aurora* entering Grand Harbour Malta for an official visit, June 1943.

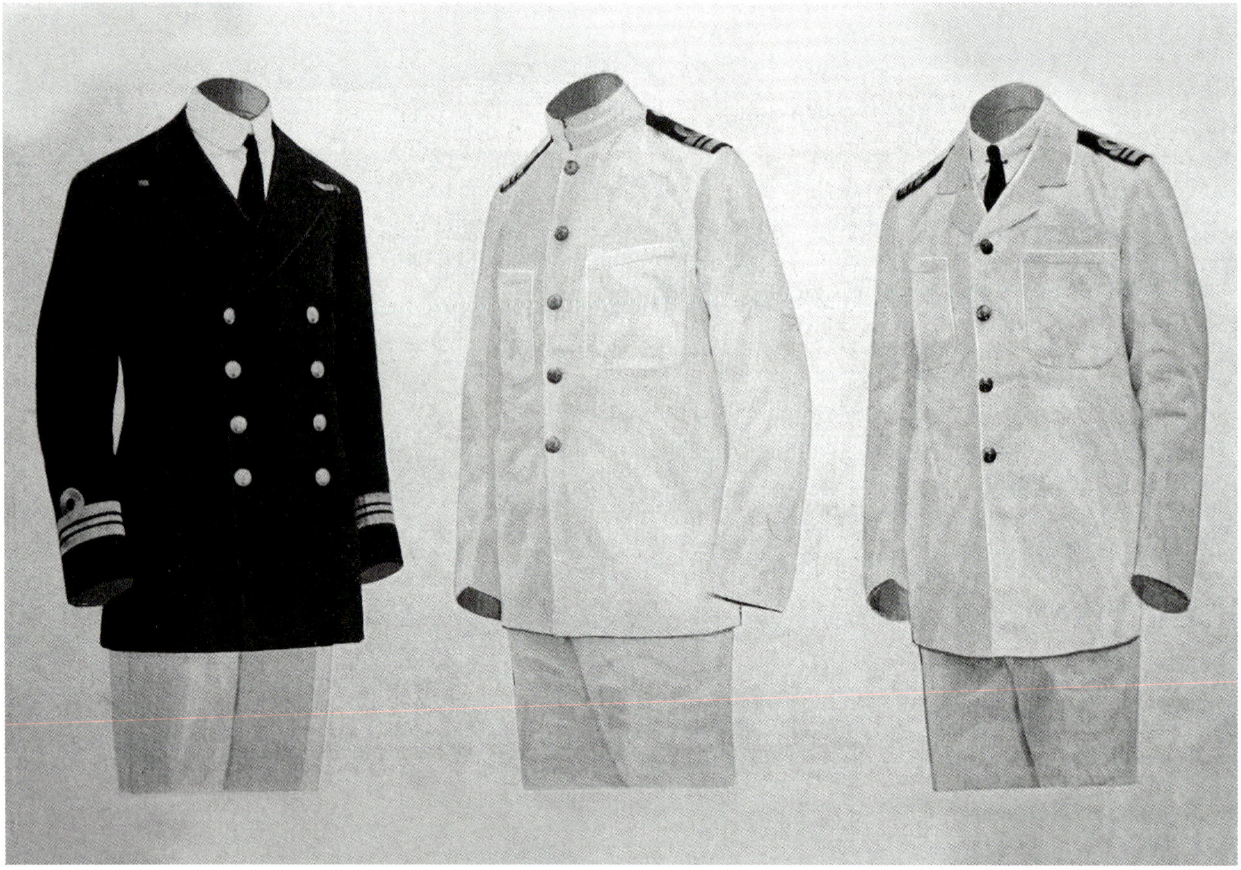

Officers' tunics as illustrated in Uniform Regulations. *At left* is the Undress Coat, *centre* is the White Tunic and *at right* the White Coat. On ceremonial occasions the wearing of the White Coat or Tunic was at the discretion of the senior Officer, at other times it was left to personal choice. BR81 1924

White Tunic

The White Tunic was a single-breasted cotton tunic with stand collar and five-button front. It had an open top patch pocket at each breast, and fittings for shoulder straps at each shoulder. The gilt buttons were removable for laundering.

The White Tunic was worn with No. 8, No. 9 and No. 10 Dress.

The white tunic worn as White Undress (No. 10 Dress) by a Royal Navy Sub Lieutenant. The tunic had five buttons to the front with rank straps at the shoulders. The dress is worn with the cap with white cover and buckskin shoes.

Detail of the RN Sub Lieutenant rank board as worn on the White Undress uniform. The strap laces through eyelets woven on the shoulders.

An inter-war photograph showing Officers wearing a mix of the White Tunic and White Coat as what was then the 8B Dress (White Undress).

White Coat

The White Coat was a single-breasted tunic similar to the White Tunic but with a step collar and four buttons. It was to be worn with a white shirt and black tie. On ceremonial occasions wearing of the White Coat or Tunic was at the discretion of the senior Officer, at other times it was left to personal choice. In early 1931 it was announced that the White Coat was not to be worn after 1932. However, under AFO 2612/31 issued in November, the abolition of the White Coat was postponed until 1 January 1935, when it was finally to be removed from the *Dress Regulations*. In the interim period the coat could be worn by those Officers who possessed one, but no new purchases were allowed.

It was not until 1950 that an open-neck Tropical Jacket once again became an item of RN clothing. Under AFO 41/1950 an open-neck Bush Jacket was introduced for Officers of Captain and above as an optional item of dress to be worn instead of the Tropical Shirt, or Tropical Tunic where appropriate.

The White Coat was worn with the No. 8B Dress. 8B Dress was abolished in 1932.

Rank straps for an RNVR Skipper (Warrant Officer) fitted to the shoulder of the Officers' Tropical Shirt.

Shirt, White Twill

The white twill shirt worn with the tropical No. 13 Dress was of the open-neck polo style with attached collar. It had a three-button front and open top pockets at each breast. At the rear shoulder there was a yoke with box pleat for

LEFT: Officer's Tropical Shirt for wear with No. 13 Dress. The shirt has a three-button front fastening, short sleeves and two open top breast pockets.

BELOW: An RNVR Sub Lieutenant wearing an open-neck white shirt with shorts, white stockings and buckskin shoes of the No. 13 Tropical Dress. The shirt is a common non-regulation pattern with full-front opening. The officer carries his Pilot's Log Book, RAF form 414, pressed into FAA service. The pilot or observer wings were not to be worn with No. 13 Dress, although this order was usually ignored.

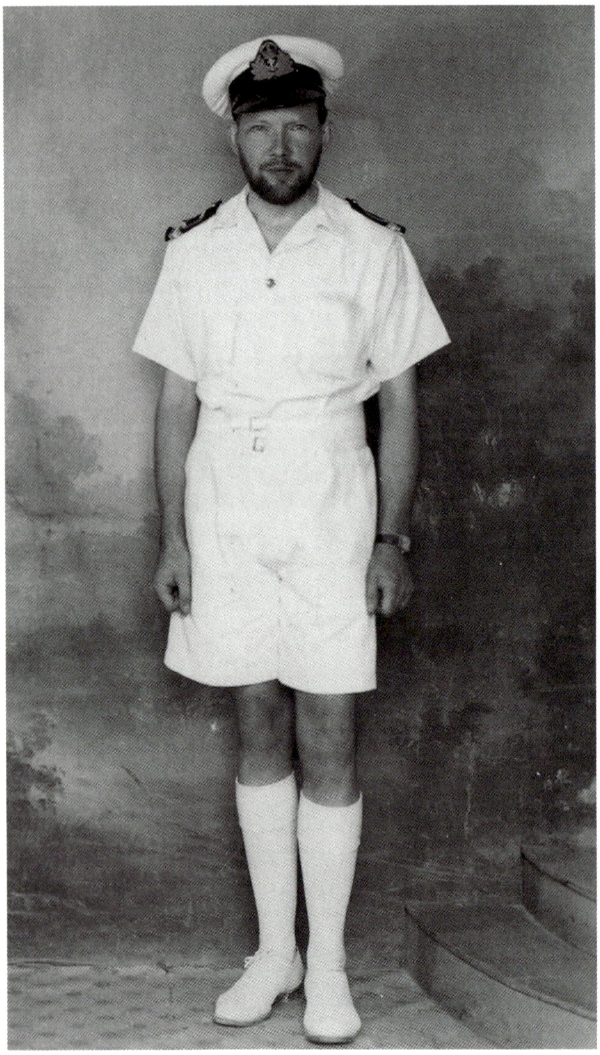

Malta, 1943. An RNVR Paymaster Lieutenant wearing No. 13 Tropical Dress. He wears the Class I and III Ratings' Tropical Shirt with rank straps and gilt button fitted to the front opening.

ABOVE: The Class I and III Ratings' Tropical Shirt fitted with Officers' shoulder straps. The use of the Ratings' shirt by Officers was commonplace. The shirt was fitted with rank straps and a plain white button, or an Officers' gilt button.

LEFT: Detail of the Sub Lieutenant's Air Branch shoulder straps laced to the shoulder of the Tropical Shirt. All Air Branch officers wore an 'A' in the curl of their rank lace. Shoulder straps were introduced into the Royal Navy in 1891.

Rank straps for a Paymaster Lieutenant RNVR as fitted to the modified Ratings' shirt. Tailored eyelets have been added to each shoulder to accept the straps' laces.

added comfort and mobility. No spine pad was fitted to the Officers' shirt. The shirts were fitted with lace holes for the fitting of shoulder straps. The sleeves could be either long or short at the discretion of the individual Officer, but short sleeves were the norm.

It had been proposed in 1937 that the Class I shirt used by CPOs be adopted for use by Officers, but the suggestion was not approved. However, during World War II it was common to see the Class I and III Tropical Shirt being worn by Officers.

Naval Sun Helmet (Officers' Uniform Helmet)

In 1881 the Commander-in-Chief China Station took it upon himself to authorize the use of white Sun Helmets with a blue pagri by Officers under his command. Of the Army 'Foreign Service Helmet' style, this was very much a local authority, and the helmet was not approved as standard by the Admiralty. The following year, in 1882, the Naval Brigade found themselves in action in the Anglo-Egyptian war. The Brigade's uniform was wholly inadequate for the parched terrain and thus the Commander-in-Chief Mediterranean authorized the use of a variety of clothing and accoutrements, including white Sun Helmets.

In 1885 the Admiralty finally approved a tropical uniform that included a white helmet for use by Officers, styled on the Army's Foreign Service Helmet, with blue pagri. At the suggestion of Admiral Lord John Hay the helmet's blue pagri was replaced by a white one with a blue fold. Throughout this period Ratings wore the simple straw Sennit Hat in hot climates.

Paymaster Midshipmen serving on HMS *Bulolo* wear the Ratings-pattern Sun Helmet. The distinctive shoulder straps worn by Paymaster Midshipmen bore the accountant branch distinction of a white cloth strip.

Officers' Sun Helmet of the Wolseley pattern introduced in 1906. It has a seven-fold pagri, with the uppermost fold being blue.

In August 1906 Admiralty Circular Letter No. 106 introduced a new pattern of Sun Helmet for naval Officers. The new Sun Helmet was of the Wolseley pattern with an improved brim giving better protection from the sun. It had a semi-pointed front peak and a broad rear brim with rounded corners. It was fitted with a seven-fold pagri – six folds of white with an inset blue fold at the top edge. The helmet had a ventilated top disc and a brown calf-leather chinstrap. The liner of the Sun Helmet was modified in 1930. The old zig-zag ventilator between the headpiece and helmet was discarded and replaced by two fibre bands.

In August 1944 AFO 4538 advised that for the duration of the war Officers were not required to provide themselves with a Sun Helmet, and that they could wear a white cap or cap with white cover on all occasions for which regulations prescribed the Sun Helmet. Officers whose duties necessitated the wearing of a Sun Helmet for protection against the sun could, if required, be supplied with a helmet on loan from the stock held for loan issue to the ship's company.

The Sun Helmet was always worn with No. 8 White Full Dress, and could be authorized for wear with No. 9 White Dress and No. 10 White Undress at the discretion of the senior Officer, and optionally with No. 13 Tropical Dress.

Action Working Dress

Action Working Dress was worn as No. 14 Dress. It is described in detail in the miscellaneous section.

Officers' Jackets, Cloaks and Coats

Mess Jacket, White Mess Jacket

The Mess Jacket was of blue cloth with padded turn-down collar. It was double-breasted with two rows of three buttons, none of which buttoned, and two rows of four ornamental buttonholes in each lapel and two pairs of buttonholes below. The front of the jacket was closed by a two-button link fitting positioned on the button holes second from bottom. It had two welted pockets, one at each side. Rank lace and coloured distinction cloth fill was worn at the cuff, according to rank and branch. The shoulders had fittings for the use of epaulettes, except for Mess Jackets worn by Warrant Officers who did not wear epaulettes. Mess Dress was not worn by Midshipmen and Cadets, who instead wore the Round Jacket.

The White Mess Jacket was similar to the Mess Jacket, but was cut shorter and had no ornamental buttonholes on the lapel. Epaulettes were not worn with the White Mess Jacket, only shoulder straps.

A Lieutenant Commander RN wearing Mess Undress uniform of Mess Jacket and Evening Waistcoat. This was the everyday dinner dress of officers.

The White Mess Undress uniform worn by a Lieutenant Commander RNVR The White Mess Undress uniform consisted of the white Mess Jacket and Evening Blue Waistcoat, or the Kamarband as worn here. This was the daily evening dress in tropical climates.

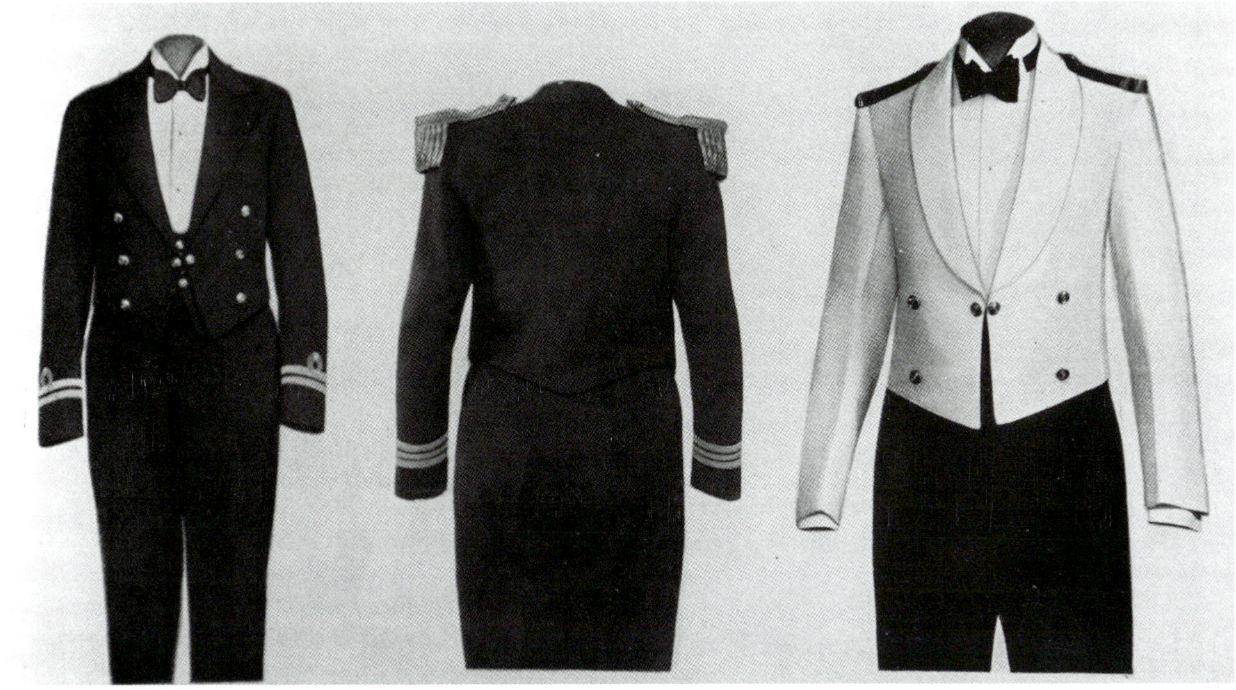

Officer's Blue and White Mess Jackets as illustrated in uniform regulations. BR81 1937

The Blue Mess Jacket as worn with No. 2 and No. 2A, No. 6 and No. 7 Dress. This example is laced to a Sub Lieutenant engineer Royal Navy. It is fitted with epaulettes for Ball Dress.

BELOW: Detail of the White Mess Jacket as worn with No. 11 and No. 12 Dress. This example is laced to a Lieutenant Commander RNVR post August 1942.

The Round Jacket and blue Evening Waistcoat combination worn by Cadets and Midshipmen as evening wear. BR81 1937

In 1942 the requirement for Mess Undress, Dress No. 7, was deleted from the initial outfit required by all temporary Officers and Warrant Officers. It was to be retained by permanent commissioned Officers. AFO 107/42 officially approved the retention of White Mess Undress No. 12 in the initial outfit of all Officers purchasing tropical rig, but deleted the walking stick from kit requirements. Officers not in possession of Mess Dress frequently wore the unauthorized mix of the Undress Coat and bow tie during the war.

In May of 1946 the use of the bow tie with the Undress Coat as Mess Dress was officially recognized.

The Undress Tail Coat was an optional item of dress that could be worn by Flag Officers, Commodores and Captains as an alternative to the blue Mess Jacket with No. 2, 6 and 7 Dress. Here it is as worn as No. 2 Dress. BR81 1937

Undress Tail Coat

The Undress Tail Coat was an optional item of dress that could be worn by Flag Officers, Commodores and Captains as evening wear. It was a double-breasted blue cloth coat similar to the Mess Jacket but with long tails. It had two rows of three buttons, none of which buttoned, and two rows of six buttonholes, four in each lapel and two below. The front of the jacket could be closed at the wearer's discretion, by a two button link. Rank lace and coloured distinction cloth fill were worn at the cuff, according to rank and branch. The shoulders had fittings for the use of epaulettes. The Undress Tail Coat could be worn instead of the Mess Jacket with Nos 2, 6 and 7 Dresses.

Waistcoats and Kamarbands

Officers were required to posses three patterns of waistcoat. The Morning Waistcoat was a blue wool waistcoat with six gilt button front, a pocket at each lower front and a pocket at the left breast. It was worn for extra warmth when wearing the Undress Coat with No. 5 Dress. The Evening White and Evening Blue waistcoats were worn with mess dress. The Evening White mess waistcoat was backless. A Kamarband (cummerbund) of black silk could be worn with No. 12 White Mess Undress uniform instead of the Evening Blue waistcoat.

The Morning Waistcoat worn with the Undress uniform for added warmth.

The blue and white evening waistcoats worn as evening wear with the Mess Jacket. The white waistcoat has a front section only with a waist and neck strap.

Officers' waistcoats as illustrated in uniform regulations: Morning, Evening White and Evening Blue. BR81 1937

Boat Cloak

An optional item of clothing, the Boat Cloak was a simple armless cloak with a lion-head neck chain fastening. The cloak was cut so that it could be worn over a coat or jacket fitted with epaulettes. As an item of dress the cloak was usually worn over the Mess Jacket for formal dinners and similar events. The cloak was one of the items of dress that was landed under AFO 2605/39.

The Officers' optional Boat Cloak. *At left* is the combined coat and cape, and *at right* the long cape to be worn without coat. BR81 1937

Watch Coat (Watchcoat)

Based on the 'British Warm' worn by the Army, the Watch Coat was a fine quality blue beaver cloth coat with a wool lining. From the front the Watch Coat looked very similar to the Great Coat. It was double-breasted with two rows of buttons, four of which buttoned. The collar was normally worn open, but could be closed when required. There was a small handkerchief pocket on the outer left breast and two pockets with flaps at the lower front. The rear of the coat was plain with no half belt and a small 14in vent pleat. Rank shoulder straps were worn at the shoulders, fully stitched down.

The purchase of the Watch Coat by Officers was optional. However, its use was compulsory for naval cadets at Dartmouth. Upon being drafted to a seagoing ship, cadets were required to purchase a Great Coat. From 1923 the Watch Coat was supposed to be fitted with a slit for the sword hilt. However, this was frequently not added to Watch Coats.

AFO 3673 issued in November 1939 stated that for the period of the war Officers were allowed to wear the Watch Coat either in addition to, or instead of, the Great Coat.

HM King George VI greets Officers following the presentation of awards. The King wears the Watch Coat.

A period advert for a life-saving coat based on the Watch Coat. The amazing properties advertised included the following: it was waterproof, gale-proof, flameproof, four times warmer than any other woollen coat at a quarter of the weight, it had six times the buoyancy of cork, and it was guaranteed to float the wearer upright in water for at least seventy-two hours. All for a mere six guineas posted.

The Watch Coat worn by a Captain RN. It is worn with brown gloves and blue cap with embroidered peak.

Great Coat (Greatcoat)

The Great Coat was a long, double-breasted blue cloth coat with two rows of six buttons. The coat was to come to 14in from the ground, providing uniformity. At the rear there was a half belt fastened with two regulation buttons and a full length box pleat. A 3in vent in the left side allowed the sword belt and sword to be worn beneath the Great Coat with the sword's hilt protruding. Two pockets with flaps were positioned at the lower front. Rank shoulder straps were worn at the shoulders, stitched down.

February 1940. Captain C. H. L. Woodhouse of HMS *Ajax* and Captain F. S. Bell of HMS *Exeter* parade through London with their ships' companies following the Battle of the River Plate. Both men wear the Greatcoat with respirator haversacks slung over their shoulder.

The Greatcoat worn by an RNVR Lieutenant (Special Branch). The Lieutenant carries the standard brown gloves and service respirator slung over the right shoulder.

The Waterproof Coat as worn by Officers, a particularly shapeless garment. No rank was visible when the raincoat was worn.

Waterproof Coat

The Waterproof Coat worn by Officers was a single-breasted fly-fronted garment. It was somewhat shapeless, and did not bear shoulder straps so when in use the wearer's rank was not visible. It was worn to 5in below the knee to provide maximum protection.

Gaiters

Officers' gaiters were made from black enamelled hide with four eyelet holes and leather laces and a calf strap with buckle fastening. They were approximately 10in high at the front and 11in at the rear. The gaiters were to be worn only in muddy or wet weather with Undress uniform at the discretion of the senior Officer. The gaiters were also worn in white uniform by Officers operating with landing parties.

In 1921 'Coastguard Pattern' black leather gaiters were introduced for optional wear by all Ratings, but only while on leave. Victualling Ratings and Officers' stewards and cooks were allowed to wear the gaiters on duty ashore in wet weather.

Officers' gaiters were made from enamelled leather, giving them a high gloss finish. They were worn in muddy or wet weather with Undress uniform and in white uniform with landing parties and by officers in charge of men under arms. JOHN F. LAMONT

HM King George VI inspects a Royal Guard in the shadow of HMS *Victory*. The Guard Commander, with drawn sword, wears leather gaiters.

Headgear

Cocked Hat

The Cocked Hat consisted of a crown section and two side flaps: the left-hand flap was 6in high and the right 5½in, and both were bound in a black silk oak-leaf braid or gold braid for Flag Officers and Commodores First Class. Positioned on the right flap and fitted over a silk cockade was a loop of bullion with button at the lower edge, the style of loop varying according to rank. Warrant Officers wore no bullion loop, just the cockade. At the front and rear points a gilt and blue tassel was fitted. It was not used by Midshipmen and Cadets: the cocked hat could also be worn with the Great Coat at funerals in bad weather.

The Cocked Hat with single bullion loop and button worn by Commissioned Officers from Warrant Rank, Sub Lieutenants, Lieutenants and Lieutenant Commanders.

Officers' Caps

The Blue Cap had a circular crown of between 9$^{13}/_{16}$in and 11$^{5}/_{16}$in, depending on size, with a piped edge. The cap was fitted with a leather chinstrap and a mohair band of 1¾in. The cap badge was fitted to the mohair band. The White Cap was similar to the Blue Cap but with a crown and quarterings of white horsehair and without edge piping to the crown.

The peak of Flag Officers' and Commodores First Class was covered with blue cloth, bound with patent leather and embroidered all round with gold oak leaves. The peak ornamentation of the cap worn by Commodores Second Class, Captains and Commanders was similar but was only embroidered on the front edge. The peak of caps worn by all other officers was of plain black patent leather. During World War II the peak depth was 2in, but in 1946 this was increased to 2¼in for caps worn by Lieutenant Commanders and below, and to 2⅜in for the braided peaks of senior Officers' caps.

Officers' cap types. *At left* a blue cap with Khaki Drill cover for use when dressed in Khaki Drill uniform, and *at right* a blue cap with white summer cover. *Above* is a blue cap with cloth-covered peak and oak leaves as worn by Commodores Second Class, Captains and Commanders.

The Commanding Officer of 1820 NAS photographed while serving at US Naval Air Station Brunswick during 1944. The Commander wears a US Navy khaki shirt and tie with RN Blue Cap with khaki cover. Use of USN kit was unauthorized but common among Officers and Ratings serving in the Continental USA.

A Sub Lieutenant RNR takes a compass bearing as his vessel enters Tobruk harbour. He wears the standard Blue Cap with white cover.

The ribbed white pique cap cover or White Cap was always worn with white trousers or tropical clothing, and with blue trousers at home from 1 May to 30 September inclusive. It was worn abroad at the discretion of the senior Officer. AFO 1307 of April 1940 instructed Officers that White Caps or White Cap covers were not to be worn ashore in the United Kingdom or on board ships in home waters for the duration of the war.

A cap cover of Khaki Drill for wear with khaki clothing was introduced during the war. It was an optional item to be worn with Khaki Drill clothing, but it was also occasionally used with Battledress.

Beret, Working Dress, Air Crews

In May 1945 AFO 2424 approved a Navy blue Beret for use with Working Dress 5B and 3A, issued to aircrews. The AFO also authorized the use of the CPO and PO Sun Helmet badges pattern 53 and 54 with the beret. Other Ratings were to wear a flash bearing the letters H.M.S. Officers were to wear a metal or embroidered badge of a similar design to the regulation cap badge. The AFO advised that the berets, flashes and Officers' badges were being prepared, and that a further AFO would give details of when the articles were available. The berets and badges or flash were to be added to the compulsory kit of all Rating pilots, observers and telegraphist air gunners. Officers would be able to purchase berets and badges from service stocks.

The officer's 'Beret, Working Dress, Air Crews' approved in May 1945 for wear with the 5B dress. This is a fine quality private purchase example with a fabric band.

Above is the small two-thirds-sized beret badge pattern 50 for use with the 'Beret, Working Dress, Air Crews' introduced in 1945. Below left is the standard gold wire-embroidered 'Badge, Cap, Pattern 47', and at right is the 'Badge, Cap, Pattern 47 (metal)' introduced in 1941 as a war economy measure.

The Commander of HMS *St Vincent*, D. L. Neame, photographed in 1938. He wears Undress uniform with the winter cap with peak embroidery as worn by Commodores Second Class, Captains and Commanders.

Officers' Cap Badges

The first cap badge authorized for the use of Officers and Warrant Officers fleet-wide was introduced in 1856. It consisted of a silver fouled anchor surrounded by a laurel wreath and surmounted by a crown. The design underwent a number of minor changes over the years with a new design approved in 1920 being the pattern that was to see service during World War II.

The Officers' pattern 47 cap badge was an ornate, hand-embroidered, gold wire device that was worn only on the Officers' peaked cap. The intricate design and the massive influx of Officers to the Navy during the latter part of 1939 led to a severe shortage of the badge. The gold wire used in the badge made it expensive, and the hand-embroidered design made production time-consuming. To help alleviate shortages, the provision of a metal version of the badge was announced in 1941, with stocks being available by the spring of 1942. When the metal badge was first announced it was called the 'Badge, Cap, Pattern X47'.

In 1945 the issue of a black beret for use by air crews saw the introduction of a special small size 'Beret badge, Officers (metal)'. This was a reduced scale, two-thirds-sized version of the standard pattern 47 (metal) badge. The badge was given the pattern number 50.

Accessories

Sword

The standard Officers' sword dated to a pattern introduced in 1827. The 1827 pattern sword was modified in 1929, having a lighter straight blade of 31¼in with a spear point. The mounts were of gilt with a solid half-basket guard embossed with a crown and anchor badge. The backpiece was finished with an ornate lion's head. The grips were of white fish skin bound with three gilt wires. The scabbard was of black leather with gilt chape and lockets. The chape was 7½in long on scabbards for Officers of flag rank and Commodores First Class, and 6½in for other Officers. For Flag Officers and Commodores First Class the scabbard

ABOVE: Detail of the hilt and the blue and gold wire sword knot of the 1827 pattern RN Officers' sword. It has a lion head back piece and an anchor and crown device on the guard. The scabbard locket has fluted threads and scrolls engraved.

LEFT: The 1827 pattern RN Officers' sword. This is a post-1929 example with the lighter, straight blade introduced in that year.

fittings were engraved with embossed oak leaves, for other Officers the fittings were engraved with fluted threads and scrolls.

The sword was worn with a sword knot fitted to the guard. The knot was a two-piece design of double-stranded gilt and blue cord and a barrel-shaped end fitting.

Full Dress Sword Belt

The Full Dress belt was made of leather with a black silk webbing. The girdle section was 1½in wide with 1in slings. Both girdle and slings were embroidered with gold wire ornamentation, the style being dependent on rank. An inner line of acorns and oak leaves with a plain edge band

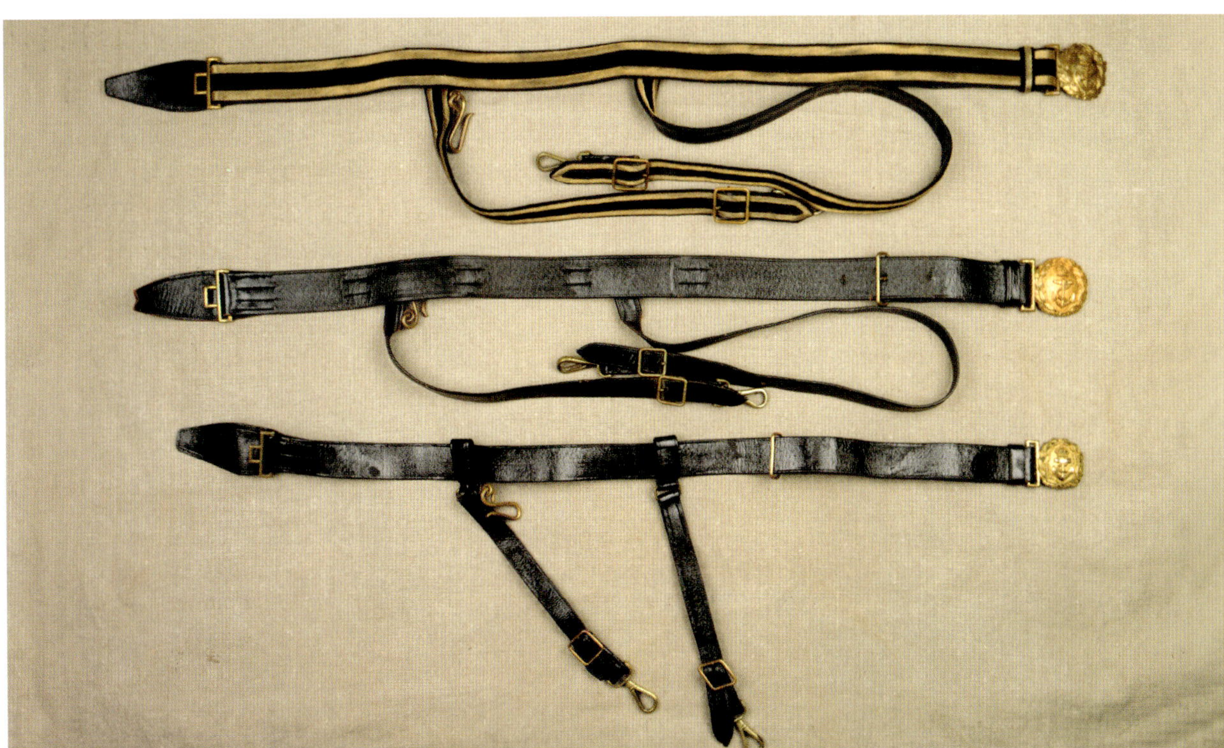

Officers' belts. *Top* is a Full Dress Sword Belt with two lines of gold embroidery as worn by Sub Lieutenants, Lieutenants and Lieutenant Commanders. *Centre* is the standard Undress Sword Belt for all officers of the rank of Sub Lieutenant and above, *below* is the Dirk Belt worn by Midshipmen with the Dirk.

A wartime advertisement for the bespoke tailors Gieves, showing Dress uniform items including Cocked Hat and sword. Prior to 1940 Gieves had been located at 21 Old Bond Street in London but were bombed out and moved to 80 Piccadilly. The firm relocated to 27 Old Bond Street in 1946. Addresses on tailors' labels can be most useful in dating garments and accoutrements.

Gilt belt clasp used with the Dress and Undress belts for ranks below Flag Officer (above), and the smaller buckle of the Midshipman's belt (below). Flag ranks used a similar buckle but with an additional outer circle of laurel.

was used for Flag Officers and Commodores First Class. Commodores Second Class, Captains and Commanders had three horizontal lines of plain embroidery, one at each edge and one in the centre. Lieutenant Commanders, Lieutenants and Sub Lieutenants had two lines of embroidery. Full Dress sword belts were not worn by Commissioned Officers from Warrant rank or by Warrant Officers. The use of the Full Dress sword belt was discontinued in 1951.

Undress Sword Belt

Both the Full Dress and Undress sword belts were modified in August 1932 under AFO 1987. The new belt used a modified attachment for the suspension slings, with each sling being fitted to a slider on the inside of the belt. The old pattern had used a brass D ring for attachment of the front sling and a sliding runner for the rear sling. Amendment No. 4 to *Uniform Regulations 1924* required that Undress sword belts were to be made of seal and not Morocco leather. Morocco leather had become unobtainable during the Great War and seal had been substituted. However, it was found that the seal had a superior finish to Morocco leather and it was thus decided to make the substitution permanent.

The Undress sword belt was worn over the Full Dress and Frock Coat, between the two lowest buttons, but under the Undress Coat and White Tunic. When worn with Ball Dress it was worn under the waistcoat. The girdle was 1½in wide with ¾in slings.

The buckle of Full Dress and Dress belts was the same. For Flag Officers and Commodores First Class it consisted of an anchor surmounted by a crown within a laurel wreath, the whole surrounded by a circular wreath of laurel. For Officers of Commodore Second Class and below it consisted of an anchor surmounted by a crown within a laurel wreath. In February 1935 a new buckle was approved for the Dress and Undress belts of Officers of Commodore Second Class and below. The general design was similar to the preceding pattern but had an anchor with improved proportions and improved detail in the anchor, crown and wreath.

Midshipman's Dirk and Belt

Introduced in 1856, the Midshipman's dirk was worn by Midshipmen, Paymaster Midshipmen, Cadets and Paymaster Cadets. It was not required to be held while serving at Dartmouth, but it was needed when drafted to a seagoing ship. It had a straight gilt and blue embossed blade

The Midshipman's Dirk as worn by Midshipmen, Paymaster Midshipmen, Cadets and Paymaster Cadets. It was not worn by Midshipmen of the Engineering branch. DIRK COURTESY ANTIQUES STOREHOUSE

of 17¾in, and a 5¼in hilt with lion's head backpiece and white fish-skin grips. The crossbar had a crown and anchor device and a spring clip to hold the blade in the scabbard. The scabbard was of black leather with gilt fittings, and was suspended by two rings fitted to the locket. The dirk knot was a reduced size version of the sword knot. The dirk belt was styled on the Undress sword belt, but with a width of 1⅛in and a smaller 1¾in clasp. The slings were fitted to two looped runners that fitted over the belt. The Midshipman's dirk and belt was not worn by engineering Midshipmen (E).

AFO 3577/41 announced that for the duration of hostilities, the wearing of swords by Sub Lieutenants and Warrant Officers and dirks by Midshipmen was to be discontinued. By default this order removed the requirement for these Officers to hold the Undress sword or dirk belt. The dirk and belt remained in abeyance until their use was finally discontinued in 1951.

Officers' Rank Lace (Distinction Lace)

Officer's cuff rank lace had seen use by Captains and Commanders between 1748 and 1767. By 1856 all Naval Officers were wearing cuff lace to show their rank. Executive (military) branch Officers had a curl on the uppermost lace stripe, while the civil branches had plain lace with no curl. In 1915 the executive branch curl was authorized for use by the engineer branch, and in 1918 its use was extended to all branches of the Royal Navy. In 1863 the four civil branches of navigator, surgeon, accountant and engineer were given a distinctive coloured fill between their rank lace stripes, inset below the rank lace when only a single lace stripe was worn, that would identify their role. As new branches were introduced into naval service, more colours were added.

Cuff rank lace, with distinction colours as appropriate, was worn on the Full Dress Coat (No. 1 Dress), Frock Coat (No. 3 and No. 4 Dress), the Undress Coat (No. 5 Dress) and the Mess Jacket (No. 2, No. 2A, No. 6 and No. 7 Dress). Rank was displayed on the shoulder straps of No. 8, No. 9, No. 10, No. 11, No. 12 and No. 13 Dress. Shoulder-strap rank was also used with the Great Coat, Watch Coat, No. 5A and No. 5B uniforms, as well as on battledress and KD clothing.

AFO 736/55 discontinued the use of distinctive branch colours, with the exception of Medical, Dental, Wardmaster and RNVR Special Branch. Officers were allowed to retain the colours until 1 January 1957. The RNVR Special Branch distinction was discontinued in 1959.

Prior to 1931, RN rank lace was ½in wide, but it was increased to 9/16in that year. In August 1941 it was decided that as a temporary austerity measure, cuff rank lace should

Civil branch distinction colours introduced in 1863

Navigator, pale blue; discontinued in 1867.
Surgeon (Medical), scarlet, 1863 to date.
Accountants (Paymasters), white, 1863 to 1956.
Engineers, purple, 1863 to 1956.

Civil branch additions post-1863

Constructor (Shipwright), silver grey, 1884 to 1956.
Naval Instructor (Schoolmaster), light blue, 1879 to 1956.
Wardmaster, red, 1911 to 1918, maroon from 1918 to 1951 (salmon pink 1951 to date).
Electrical, green, 1918 to 1956.
Armourer (Ordnance), dark blue, 1918 to 1951.
RNVR Special Branch, dark green, 1923 to 1959.
Dental Officer, orange 1918 to date.

only be used on the outer face of the cuff, with the inner face being void of lace. This greatly reduced the quantity of gold rank lace being used, but still allowed the rank to be easily seen and identified. Typically AFO 3345/41 authorizing the use of half cuff lace was routinely disregarded, and Officers continued to purchase uniforms with full rank lace, exacerbating the shortage. The use of full cuff rank lace was restored in 1947.

AFO 3744 of August 1942 stated that in future the half stripe of RNVR Lieutenant Commanders' cuff and shoulder strap distinction rank lace was to be waved in the same way as the broad stripe. It was not to exceed ⅛in in width. Prior to this the half stripe had been straight.

Royal Navy rank lace was straight-edged. RNR lace was double-stranded, giving the appearance of intertwined chain link. RNR Officers were drawn mainly from the Merchant Navy and the nation's fishermen. The lace worn by Officers of the RNVR had a waved shape leading to RNVR Officers being called 'Wavy Navy'. RNVR Officers were civilians with no maritime connections, who trained to assist the RN.

Midshipmen and cadets did not wear rank lace. Midshipmen wore the turnback and button on the collar, Cadets wore the notched hole with white twist and button. Paymaster Midshipmen and Cadets wore the white branch distinction colour on the cuff. RN Midshipmen and Cadets wore white turnbacks or notched hole, RNR wore blue and RNVR wore maroon.

ABOVE LEFT: Cuff rank distinction colours. Shown here are: *top left* the scarlet fill of a Surgeon Lieutenant RNVR, *top right* blue for Ordnance Lieutenant RN, *lower left* maroon of a Commissioned Wardmaster RN, and bottom right the purple of a Sub Lieutenant Engineer RNVR. The colour differences between some branch distinction lace could be quite subtle.

ABOVE RIGHT: *Above* is the blue turnback with notched hole of a Midshipman RNR fitted to the Undress Coat. *Below* is the blue notched hole of a Cadet RNR.

LEFT: Officers' rank: *Top left*, Warrant Officer Paymaster RN (white distinction), Sub Lieutenant Engineer RNR (purple distinction), Lieutenant Dental RNVR (orange distinction). *Second row*, Lieutenant Commander Electrical RN (green distinction), Surgeon Commander RN (scarlet distinction), and Captain RNR (executive branch).

ABOVE: Typical Officers' shoulder-strap types. *Top, left to right,* Lieutenant Commander RNR leather-backed Greatcoat strap (the fabric swatch retains the button); Lieutenant Commander Constructor RN, fabric backed Greatcoat strap; Lieutenant Commander Engineer RN, woven 5A Working Dress strap. *Bottom, left to right,* Lieutenant Special Branch RNVR, leather-backed White Dress/tropical laced strap; Lieutenant RNVR KD slip-on strap; Lieutenant Commander RNVR KD laced strap with bronzed button for Khaki SD jacket.

LEFT: An advertisement for Gieves Royal Naval outfitters showing the Undress Coat and variants in the peak ornamentation of the blue cap. Top is the cap worn by Flag Officers and Commodores First Class, centre is the cap for Commodores Second Class, Captains and Commanders. Below is the cap for all other Officers.

Buttons

Officers' uniforms were fitted with fine gilt buttons, the size depending on the garment to which they were fitted. The larger size was $\frac{9}{10}$in (thirty-seven lines) and for use on coats and the Cocked Hat. The intermediate size was $\frac{15}{20}$in (thirty lines) for use on jackets and the slash of the Dress Coat. The small buttons $\frac{13}{20}$in (twenty-six lines) were for use on waistcoats, epaulettes and shoulder straps. Buttons were of two patterns: for Flag Officers and Commodores First Class the button had a rope rim encircling a plain rim, within which was a laurel wreath surrounding an anchor and cable surmounted by a crown. For other Officers the button had a rope rim encircling a plain rim, within which was an anchor and cable surmounted by a crown.

In the spring of 1944 it was reported that gilt-finished steel buttons had been fitted to some uniform suits manufactured for Royal Naval Officers and purchased in the USA and Canada. It had been found that the buttons were manufactured by the Waterbury Button Co. of Connecticut and that they were capable of deflecting a magnetic compass by as much as 4 degrees. All Officers, particularly those who had purchased uniforms in the USA or Canada, were ordered to check their uniform buttons to ensure that they were not magnetic. Any found to be so were to be replaced immediately.

The USA proved to be another source of problem for the buttons. It was reported in the spring of 1945 that gold rank lace, gold wire badges and gilt buttons were suffering severe surface deterioration when worn in the US port of Philadelphia. This was due to the extremely high content of hydrogen sulphide gas in the atmosphere during the summer months. It was suggested that, as a partial remedy for any ships proceeding to Philadelphia, all gold lace, badges and buttons should be coated in banana oil or nail varnish to provide a semi-retardant finish.

Officers' button types $^{15}/_{20}$ in (thirty lines). *At left* is the standard pattern worn by all Officers from Warrant Officers through to Commodore Second Class; *at right* the ornate pattern worn by Flag officers from Commodore First Class through to Admiral of the Fleet.

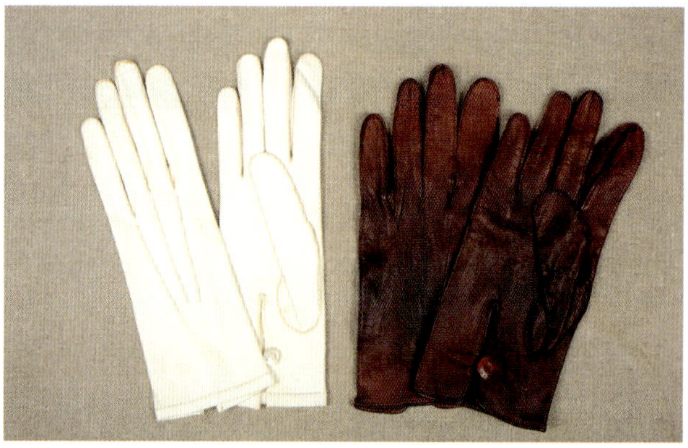

Officers' leather gloves. *At left*, white kid gloves worn with Full Dress, Ball Dress, Frock Coat with Epaulettes Dress, Frock Coat Dress, White Full and White Dresses or any other dresses except Undress. Gloves were not to be worn with White Undress. The brown leather gloves were worn with Undress uniform and could also be purchased by CPOs and POs. In cold weather white knitted gloves could be worn.

ABOVE: Officers' black leather lace-front shoes worn during the war by a Surgeon Lieutenant RNR. Naval shoes had a plain toecap.

LEFT: An Engineering Officer manning the engine room telegraph. Officers wore brown or blue overalls when the nature of their work required their use.

2 Ratings' Uniforms

Uniform and Kit Review

In 1931 a committee was established to review the uniform and kit of naval Ratings. Presided over by Rear Admiral W. F. French CMG, it was to consider all aspects of uniform, the quantity issued, the quality, and its suitability for the intended task. In 1932 the results of the review were published in AFOs. On the whole the recommendations were concerned with the quantity of items issued, with the compulsory possession of some items of dress being increased, and others decreased. The alterations came into effect as of July 1932, and the men were given until January 1934 to purchase any additional compulsory items required by the changes.

A number of new and modified patterns of uniform were also adopted as a result of the review. A new tropical shirt with open neck and roll collar was introduced for CPOs to replace the tropical singlet. A new waterproof coat and lightweight peaked cap were to be provided for Classes I and III Ratings, and for Class II Ratings the jumper and trousers were to be modified to provide a better fit. To accommodate these changes the blue jean collar and flannel were also to be modified to make them suitable for wear with the new pattern of jumper. A new style of Class II blue cloth cap was also introduced. Details of the changes to individual items are listed under the entries for those items.

Kit Upkeep Allowance

Prior to December 1912 Ratings had received no free kit, but instead had been provided with a gratuity towards the upkeep of their kit, paid quarterly. From December 1912 a free initial kit issue was provided, with further upkeep being at the Ratings' expense. In 1917 'Kit Upkeep Allowance' (KUA) was granted to be credited to all Ratings: £7 for CPOs, £6 for men not dressed as seamen, and £5.10s for men dressed as seamen. Purchases of clothing were debited from the Ratings' pay from the Slops (cash clothing sales) ledger. In 1922 a new system was introduced whereby men would pay for any Slops purchases with cash, as from 1 January 1923. By 1939 the levels of KUA had been reduced, typical rates being £6.19s for CPOs and confirmed POs, £6.12s for men not dressed as seamen, and £5.8s for men dressed as seamen. An additional 'Submariner's Kit Allowance' payment of 14s was made to submariner seamen, and signal and telegraphist Ratings, while other Ratings received 11s.

Classification and Orders of Dress of Royal Navy Ratings' Uniforms

The following numbered classes of Royal Navy Ratings' uniform relate to the orders of dress being worn in June 1939, and the occasions on which they would be worn.

Class I Uniform

Class I uniforms were worn by Chief Petty Officers of all trade groups. Collectively, uniform groups Classes I and III were classed as 'men not dressed as seamen'.

Class I Dress No. 1: For inspections, musters, ceremonial occasions, Sundays in harbour and when going on leave, the No. 1 Dress was worn by CPOs. It consisted of a double-breasted cloth suit with gold badges and medals.

Class I Dress No. 2: The No. 2 Dress was worn for Sundays at sea, and it was optional clothing for weekday wear when on leave. All CPOs on duty were to wear the No. 2 Dress when the ship's company were in No. 3 Dress. CPOs' No. 2 Dress consisted of a double-breasted cloth suit with gold badges and medal ribbons.

Class I Dress No. 3: For normal working day wear the No. 3 Dress was worn. It consisted of a single-breasted serge suit with red badges. Medal ribbons were optional with No. 3 Dress. An old double-breasted suit could be worn with red badges in lieu of the single-breasted suit.

Class I Dress No. 4: For wet weather and for normal evening wear (night clothing) the No. 4 Dress was worn. Any old cloth or serge suit with red badges could be worn as No. 4 Dress. As with No. 3 Dress, medal ribbons were optional.

Class I Dress No. 5: For hot weather the No. 5 Dress replaced the No. 3 Dress as normal working clothing. It could also be worn when No. 3 clothing was not considered suitable. It consisted of the serge suit, or if whites were being worn, a drill suit. Medal ribbons were optional.

Class I Dress No. 6: At the discretion of the senior Officer, No. 1 Dress could be replaced by the No. 6 Dress. For No. 6 Dress the white drill suit was worn with medals on foreign stations where the temperature was considered too hot for cloth or serge uniforms.

Class I Dress No. 7: In warmer climates No. 2 Dress could be replaced by the No. 7 Dress at the discretion of the senior Officer. In No. 7 Dress the white drill suit was worn without medals. Prior to 1933 No. 7 Dress had been as for No. 2 but with drill trousers, or as No. 6 without medals, as ordered.

Class I Dress No. 8: In June 1939 there was no No. 8 Dress for CPOs. No. 8 Dress had previously been as for No. 3 Dress but with drill trousers, or as No. 6 without medals, as ordered. It was discontinued in 1933.

Class I Dress No. 9: For particularly dirty work such as coaling ship, painting or refitting, and when other classes of clothing might have been spoiled, CPOs would wear the No. 9 Dress. This consisted of the blue overall suit (Combination Overalls), or No. 5 Dress as ordered.

The cloth suit with gold badges worn by a CPO shipwright. The cloth suit with gold badges was worn with No. 1 and No. 2 Dress by Class I Ratings.

Cloth tunic as worn with No. 1 and No. 2 Dress by a CPO ERA (Engine Room Artificer). Artificers did not wear any non-substantive badges on the collar; their CPO rating was indicated only by the three cuff buttons and their cap badge.

Class I Dress No. 10: In tropical stations where the use of tropical clothing was allowed, the wearing of No. 10 Dress was authorized at the discretion of the Commander-in-Chief. No. 10 Dress comprised white shirt and white shorts worn with a white cap or Sun Helmet (as ordered).

Orders of Dress Introduced during World War II

Class 1 Dress 3A: Working Dress for Air Crews was to be worn by Air Crews at sea, and in aircraft at all times; also in ships in harbour, in naval and dockyard establishments, on Naval Air Stations, and Naval Air Sections during working hours or on non-ceremonial occasions.

Class I Dress No. 8: Action Dress was to be worn at action stations in hot climates, when ordered by the senior officer, in conjunction (if necessary) with other clothing. Also on shore as an anti-malarial dress if ordered by the senior Officer, and on all occasions for which the No. 5 Dress was authorized.

Uniform Kit List Class I for Chief Petty Officers

Compulsory Kit

One jacket, double-breasted, cloth, for dress
One jacket, single-breasted serge, for working dress
Two tunics, drill (three for CPO cooks and CPO stewards)
One waistcoat, cloth
One waistcoat, serge, or blue pullover
Two pairs trousers, cloth
One pair trousers, serge
Three pairs trousers, drill
Two caps, cloth, with peak
Three cap covers, Marcelline
Two vests, summer
Two vests, winter
Two pairs drawers
Six shirts, white
Twelve collars, white
Two neckties, black silk
Two pairs socks or stockings
Two pairs boots, half, or one pair boots half and one pair of shoes
One bed
One blanket
Two bed covers
One blue overall suit (compulsory for artificer, artisan and engine room Ratings but optional for other CPOs)
One waterproof coat
One pair braces
Three aprons (cooks only)
Three caps, cooks, working (cooks only)

Optional Kit

One jacket, double-breasted, tartan
One waistcoat, tartan
One pair trousers, tartan
One jacket double-breasted, diagonal serge
One waistcoat, diagonal serge

One pair trousers diagonal serge
One overcoat
One oilskin coat
One sou'wester
One pair oilskin leggings
One pair shoes black leather
One pair shoes, brown canvas, or slippers, black, leather
One pair shoes, white canvas
Two pairs socks, white
Two suits, pyjamas
Two belts, cholera
One pair gloves, woollen
Six handkerchiefs, white cotton
One brush, shaving
One box, cap
One pair gaiters, black leather
One suitcase
One attaché case
Two shirts, blue and white striped
Four collars, check, blue and white striped

Boarding party. A CPO gunner's mate wearing the normal daily working rig for temperate climates, the No. 3 Uniform's single-breasted serge jacket with red badges. The CPO models the 1919 web equipment pistol order with web gaiters. At his left side he carries the Pattern 1900 boarding cutlass. The cutlass was still issued during World War II. BR159 1934

A CPO air mechanic photographed on his wedding day wearing his best No. 1 uniform, the double-breasted cloth suit with gold badges.

One scarf, plain white
One pair gloves, brown leather
One pullover
One pillow
Two pillow covers
Two sheets, white
One waistbelt, white
One cover (washable) for gasmask haversack

Tropical Clothing (when ordered)
Tropical shirt with $\frac{9}{16}$in gilt crown and anchor button
White shorts
White stockings
White canvas shoes

Class III Uniform

Class III uniforms were worn by all POs and men not dressed as seamen, and men wearing 'fore and aft' rig. Collectively uniform groups Classes I and III were classed as 'men not dressed as seamen'. Other than the cuff buttons and collar badges, Class III Dress numbers were essentially the same as those for CPOs Class I, except for Class III Dress No. 2.

Class III Dress No. 1: As for Class I.

Class III Dress No. 2: The No. 2 Dress was worn for Sundays at sea and by signal Ratings on duty on Sundays when in harbour. It was optional clothing for weekday wear when on leave. All duty men, boat crews and signallers were to wear the No. 2 Dress when the ship's company were in No. 3 Dress. For POs and men not dressed as seamen, No. 2 Dress consisted of a double breasted cloth suit with gold badges and medal ribbons.

Class III Dress No. 3: As for Class I.

Class III Dress No. 4: As for Class I.

Class III Dress No. 5: As for Class I.

Class III Dress No. 6: As for Class I.

Class III Dress No. 7: As for Class I.

Class III Dress No. 8: As for Class I.

Class III Dress No. 9: As for Class I.

Class III Dress No. 10: As for Class I.

Orders of Dress Introduced during World War II
Class III Dress No. 3A: As for Class I.

Class III Dress No. 8: As for Class I.

Uniform Kit List Class III

For Petty Officers and men not dressed as seamen: June 1939.

Compulsory Kit

One jacket, double-breasted, cloth, for dress
One jacket, single-breasted, cloth, for dress (musicians only)
One waistcoat, cloth
One pair trousers, cloth
One jacket, single-breasted, serge, for working dress
One waistcoat, serge, or blue pullover
Two pairs trousers, serge
Three pairs trousers, drill
Two tunics, drill
Two caps, cloth, with peak
Three cap covers, Marcelline
Three caps, cooks, working (cooks only)
Three aprons (cooks only)
Two vests, summer
Two vests, winter
Two pairs drawers
Five shirts, white
Twelve collars, white
Two neckties, black silk
Two pairs socks or stockings
Two pairs boots, half, or one pair boots, half, and one pair of shoes
One bed
One blanket
Two bed covers
Two towels
One type, name
One soap bag
One comb, hair
One brush, hair
One brush, tooth
One brush, clothes
One brush, boot, hard
One brush, boot, polishing
Two blue overall suits (for artificer, artisan and engine room Ratings only)
One blue overall suit (supply Ratings below the rank of Petty Officer only)
One waterproof coat
One pair braces
Manual of Instruction for Sick Berth Staff (sick berth staff only)
Engineering Manual (ERAs and mechanics only)
Instructions for Central Storekeeping (supply Ratings only)
Manual of Naval Cookery (Rating cooks only)
Manual of Victualling (supply Ratings only)
Manual for Officers' Stewards (Officers' stewards only)
Handbook for Officers' Cooks (Officers' cooks only)

Optional Kit

One jacket, tartan
One waistcoat, tartan
One pair trousers, tartan
One jacket, diagonal serge
One waistcoat, diagonal serge
One pair trousers diagonal serge
One overcoat
One oilskin coat
One pair oilskin leggings
One sou'wester
One pair shoes black leather
One pair shoes, brown canvas, or slippers, black, leather
One pair shoes, white canvas
Two pairs socks, white
Two suits, pyjamas
Two belts, cholera
One pair gloves or mitts, woollen
Six handkerchiefs, white cotton
One brush, shaving
One box, cap
One pair gaiters, black leather
One suitcase
One blue overall suit
One comforter
One attaché case
Two shirts, blue and white striped
Four collars, blue and white striped
One scarf, plain white
One pair gloves, brown leather
One ditty box
One pullover

One pillow
Two pillow covers
Two sheets, white
One waistbelt, white
One cover (washable) for a gas mask haversack

Regulating PO 'Crusher' wearing the double-breasted cloth suit as No. 2 Dress. His regulating role is shown by the crown badge worn on the upper right sleeve. The RPO was responsible to the Master-at-Arms for the maintenance of good order and discipline on board.

The pattern 247 gold wire crown badge worn by Regulating POs on the right sleeve.

Tropical Clothing
Tropical clothing No. 10 Dress was not authorized for use on the Mediterranean station until 1937. Prior to this the No. 5 Dress was worn as working dress. No. 6 and No. 7 Dress were worn for more formal occasions in hot climates.

Confirmed Petty Officers
Tropical shirt with $\frac{9}{16}$in gilt crown and anchor button
White shorts
White stockings
White canvas shoes

Other Ratings not Dressed as Seamen
Tropical shirt with $\frac{9}{16}$in black horn button
White shorts
Blue stockings
Black boots or shoes

A Class III Leading Rating dressed in the white drill uniform with the distinctive horn buttons used by Acting Petty Officers and below.

Class I and Class III Uniform Items

On 1 June 1939 AFO 1475 gave details of alterations to the holdings of Ratings uniform and kit in the event of mobilization for war. In October 1939 AFO 3025 promulgated these modifications. The wearing of No. 1 Dress and gold badges was suspended for all Ratings except as optional for wear while on leave. The suspension of No. 1 Dress also meant that medals could no longer be worn. The

New entrants class 190 photographed at HMS *Royal Arthur* at Skegness. The 'Tiffs' wear shirt sleeve dress, a locally authorized hot-weather rig in non-tropical stations often called 'half blues'.

compulsory requirement for 'One jacket double-breasted, cloth, for dress, one waistcoat, cloth, and one pair trousers, cloth' was removed from Ratings uniform regulations. The double-breasted cloth suit was from that point an optional clothing item for those who had possession of them, but no new cloth suits were to be procured. An additional single-breasted serge suit was added to the compulsory kit in lieu of the double-breasted cloth suit. To avoid unnecessary waste, gold badges and suspended uniform items could be purchased from Slops until the stocks were exhausted.

The pre-war Class I and III compulsory kit holding for white drill tunic and trousers was two tunics and three pairs of trousers for all Ratings not dressed as seamen. However, for Class I and III Ratings actually serving on foreign stations where white drill clothing was worn, the compulsory level was increased to four tunics and four pairs of trousers. AFO 3025/39 reduced this holding to two tunics and two pairs of trousers.

No. 1 Dress with gold badge double-breasted suits was reinstated for all CPOs and confirmed POs serving overseas or on sea-going ships in 1951.

Class I and III button types. From left, black horn button for men and acting POs not dressed as seamen; white horn button, drill uniform for men and acting POs not dressed as seamen; gilt button for CPOs and confirmed POs.

Master-at-Arms Frock Coat

Although not ranked as Officers, the unique position of the Master-at-Arms brought with it several privileges above those of other senior Ratings. This included the use of a single-breasted Frock Coat and a sword with black grips.

In February 1940 AFO 709 advised that the wearing of the Frock Coat on dress occasions by Master-at-Arms was to be discontinued for the duration of the war. Those Frock Coats in possession of MAA serving afloat were to be landed at the first opportunity. Thereafter regulating POs promoted to Master-at-Arms were not required to purchase the Frock Coat.

Pattern 246 Master-at-Arms badge as worn on the collar of serge tunics and cuff of the drill tunic.

A Master-at-Arms wearing the special pattern, single-breasted Frock Coat worn with a sword as No. 1 Dress for inspections and divisions. This illustration was a part of a series of cigarette cards produced by Churchman in 1937, depicting the Royal Navy at work.

Class I and III Dress Jacket (Double-Breasted Jacket)

Prior to February 1932 the dress jacket worn by Class I and III Ratings had four buttons on each side, three of which fastened. The upper buttons on each side usually sat below the collar and were not visible, but they were frequently repositioned so that they sat in the notch of the collar. These superfluous upper buttons were removed from jackets in February 1932. New production had two rows of three buttons, three to fasten, spaced equally at between 3¼in to 4in, depending on the wearer's height, with the lowest button being level with the pocket. The old pattern jackets could be used until they were worn out.

The use of the double-breasted cloth dress jacket was suspended in October 1939. The single-breasted jacket was henceforth to be worn on all occasions, and was to be the authorized pattern in lieu of the double-breasted jacket. Despite the double-breasted jacket no longer being an authorized article of kit, many Class I and III Ratings continued to purchase them from trade sources, who were usually more than happy to oblige. AFOs and other orders repeatedly advised Ratings that they were not to purchase double-breasted suits or any other uniforms not of the authorized pattern, and that any such purchases would result in disciplinary action. Those men already in possession of the double-breasted jackets were allowed to continue wearing them until they were worn out.

In March 1945 AFO 1222 advised that careful consideration had been given to the re-introduction of the No. 1 Dress cloth suit and gold badges for CPOs and POs. Unfortunately the availability of suitable cloth and the strain on manufacturing facilities had meant that it was not a practical option at that time. As a concession, CPOs and POs were authorized the option to purchase double-breasted made-to-measure suits produced from standard Navy serge – 'Serge, Blue, No. 4, Pattern T76' – in lieu of a single-breasted suit. This option was not extended to other Class III Ratings.

Class III uniforms were frequently the subject of discussion within the Fleet. Class II Ratings were indignant that Ratings of an equal rank wore a style of uniform associated with CPOs and POs, whilst those who wore the Class III uniform certainly felt themselves to be above the Class II Ratings. This view was also expressed in official documents, with one comment being that the Class III uniform provided a 'visible sign that they are drawn from a class above that of the common sailor'. On a number of occasions the removal of Class III uniform and the substitution of Class II uniform for men below the rating of PO was muted.

In December 1943 a review of the Class III uniform proposed the issue of Class II uniform for all men below the rate of PO. The proposals were placed before the various Commanders-in-Chief, and while some approved the change for all Ratings, others stated that Class II should be reserved for seamen, and a few suggested that the change should be for accountant branch only and that Artificers should retain Class III uniform. But in general the proposal was received favourably, as it was felt that it would also enhance the prestige of POs. However, the Cooks and Stewards branches were notoriously difficult branches in which to maintain a satisfactory supply of recruits, and it was felt that the Class III uniform provided a 'snob value' in attracting recruits to unpopular branches that the Admiralty could not afford to neglect. It was also noted at the highest levels that the civilian unions would undoubtedly

A CPO gunner's mate puts Admiral Dreyer through his paces during a visit to HMS *Glendower.* The CPO wears No. 2 uniform, the double-breasted cloth suit with gold badges and medal ribbons. Admiral Dreyer wears the Undress uniform.

object passionately to any change to the uniform and thus the prestige and standing of Artificers and Artisans.

Ultimately it was decided that although the change would be beneficial to the service, the time was not right, and it would be unpopular and resented by many. It was to be some years before the subject was raised again. In 1956 it was finally decreed that all men below the rank of PO should in future be dressed in Class II uniform except for Artificer apprentices who were to remain dressed in Class III uniform.

The three ⅞in gilt buttons worn on both cuffs of all CPOs' tunics, both single- and double-breasted.

The Class I cloth tunic with red badges as worn by a supply CPO (stores CPO from October 1944). Old Class I cloth, tartan or diagonal serge suits unsuitable for wear as No. 1 or No. 2 Dress could be fitted with red badges and worn as No. 4 Dress night clothing.

Detail of the RPO's double-breasted cloth suit. The ribbon of the 1939–43 star is worn on the left breast.

Pattern 1 gold wire substantive rate badge for a PO, worn on the left sleeve above two pattern 42, gold wire, good conduct badges denoting eight years' good conduct, as worn with No. 1 Dress.

A cook wearing the pre-October 1931, pattern 40C, silver star badge on his double-breasted cloth jacket of the pre-February 1932 pattern. The jacket has a shallow V to the front and wide-spaced buttons, two of which are fitted beneath the lapels.

Cuff buttons on a PO's double-breasted suit. Purely decorative by the 1930s, originally they were functional fastenings on Class III double-breasted suits.

The Class III double-breasted tunic as worn by a PO Artificer with No. 1 and No. 2 Dress. Artificers did not wear a non-substantive badge. The production of the double-breasted jacket was suspended in October 1939 as a war economy measure.

A Leading Officer's Steward wearing the gold badge double-breasted cloth suit used for No. 1 and No. 2 Dress.

Class III Rating's uniform for a Leading Rate Officers' Steward. The double-breasted suit was not authorized for purchase during World War II.

Gold wire substantive badge pattern 3 for a Leading Rating worn over a single pattern 42 good conduct badge, indicating three years' good conduct.

Classes I and III Working Dress Jacket

The working dress jacket worn by all Class I and III Ratings was a single-breasted jacket with two hip pockets with flaps. CPOs and confirmed POs wore gilt buttons. CPOs wore three $\frac{7}{8}$in Rating ornamental buttons and confirmed POs wore two $\frac{9}{16}$in gilt buttons at each cuff. Other Ratings wore black horn buttons throughout. From February 1943 all ready-made and made-to-measure jackets for Class III Ratings were thereafter to be supplied and worn without the ornamental cuff buttons. Those already having cuff buttons could continue to wear them.

Pattern 138 Officers' Steward gold wire badge. Accountant branch ratings wore a star encircling the letters of the individual trade group; the star badges were introduced in 1931.

The working dress jackets were altered in May 1943. The new pattern jacket featured an improved cut, and was available in twelve sizes compared to the nine sizes of the preceding pattern. The front fastening was reduced from four buttons to three, giving a deeper

The single-breasted serge jacket with red badges as worn by a CPO cook. The serge suit was worn as working rig in No. 3 Dress. It was also worn with No. 4 and 5 Dress. The Chief is charging his pipe with 'Tickler' (duty-free tobacco).

A CPO helps a wounded Corporal of the Auxiliary Military Pioneer Corps during the evacuation from Dunkirk. The Chief wears a well used four-button, single-breasted serge tunic over a cardigan and white singlet vest.

Detail of the single-breasted serge jacket as worn by a CPO Cook. The serge jacket was used as the daily working dress.

Single-breasted Class III Working Dress jacket to a PO Submarine Detector Instructor. The three-button style replaced the four-button fastening in May 1943.

Pattern 82A badge for a Submarine Detector Instructor.

At sea the normal daily working rig of POs was the No. 3 Dress. This PO Visual Signalman 2nd Class wears the single-breasted Serge Working Jacket with red badges.

A Chief Stoker wearing the single-breasted serge tunic with red collar badges. Alongside him is a Class III rating in No. 3 Dress single-breasted jacket with black buttons, and a RNR Commander in No. 5 Dress.

Unlike Class II Ratings, Class III Ratings wore gold badges with their No. 2 Dress, a double-breasted cloth jacket with black horn buttons. This Rating is a Leading Cook with a single good conduct badge.

Gunnery CPOs and POs undergoing training at HMS *Excellent* during 1945. The men wear white drill trousers with Working Dress serge jackets, the old No. 8 Dress having been discontinued in 1933. The dress was retained as a training rig for Officers and Ratings undergoing gunnery training.

Class III single-breasted serge suits were worn with No. 3, No. 4 and No. 5 Dress. During World War II it was also worn as No. 2 Dress. This pre-May 1943 serge jacket has the four-button fastening.

Proudly stood to attention, a Class III Rating poses for the camera. He wears No. 3 Dress, a single-breasted serge jacket with red badges and black horn buttons.

A steward engaged in daily chores, maintaining the high dress standards of the ship's Officers. At action stations, stewards would form first aid and stretcher bearer parties.

Pattern 135A badge for an accountant branch writer.

Pattern 136A badge for an accountant branch supplying Rating.

A post-May 1943 Class III single-breasted Working Dress jacket with three-button front fastening.

Waterproof Coat

Introduced in 1932, the Waterproof Coat was compulsory for Class I and III Ratings, and from 1937 it was an optional item for Class II Ratings. The coat was removed from the optional kit of Class II Ratings in 1941 to reduce shortage of stock. Class II Ratings were no longer permitted to purchase the coat, but those already in possession of one could continue to wear it until worn out.

In early 1943 it was announced that the belt that had previously been supplied with the Waterproof Coat would no longer be provided once current stocks had been exhausted, and that in future the coat would be sold without the belt as an economy measure.

Waistcoat and Pullover

Waistcoats, cloth or serge, were issued to all Class I and III Ratings for wear beneath the working or dress jackets. CPOs' and POs' waistcoats were fitted with gilt buttons; other Class III Ratings used black-horn buttons. A blue wool pullover for use by men not dressed as seamen was introduced in 1938. The pullover was available as an optional item in lieu of the serge waistcoat. All Class I and III Ratings had to hold either the waistcoat or pullover as part of their compulsory kit. The pullover was a simple, plain knitted, sleeveless V-neck style, and was available in sizes 1 to 4 relating to 34in to 40in chest.

The waterproof coat was compulsory kit for all Class I and III Ratings. The mackintosh-style coat had a removable liner, fly front and waistbelt. This pattern was introduced in 1932.

A Class III Rating wearing the Waterproof Coat for a studio portrait.

LEFT: A Class III rating donning the cloth waistcoat over the white shirt. Stiff collars were standard, but soft collars were permitted with Working Dress.

The cloth waistcoat issued to Class I and III Ratings. CPOs and confirmed POs wore gilt buttons, with acting POs and other Class III Ratings wearing black horn buttons. A pullover could be worn instead of the waistcoat.

Cap, Cloth, Blue with Peak

A new pattern of blue cloth cap with a peak for Class I and III Ratings was introduced in 1923. It had an improved shape and fuller crown than the pattern worn during the Great War. It had a padded crown, leather peak and chin-strap, and a mohair band to which the cap badge was fitted. A white cap cover was worn during the summer months or with white clothing.

Cap, with Peak, Lightweight

During the late 1920s and early 1930s research was undertaken to find a lightweight peaked cap for use by Class I and III Ratings in hot climates. Such caps were available from naval outfitters and were popular in Malta. A variety of caps were provided for trials in the Mediterranean, including the pattern available in Malta, but none was found suitable for issue, frequently due to staining from sweat. The Admiralty eventually designed a cap with a crown of two layers of grey waterproof cloth, its shape maintained by a cane grommet. It had a patent leather peak of 1¾in deep, and a black mohair band. The quarterings and band were of white duck, and the head leather was perforated for ventilation. The caps, which had to be worn with a Marcelline cover, became available from victualling yards in limited numbers in 1934. They were available in nine sizes: 6⅜ to 7⅜.

Ratings who were accepted as Officer candidates and who were undergoing training for their commissions were authorized the wearing of a plain white cap band. AFO 2888 of January 1940 authorized the issue of two such ribbons to each trainee. Ratings in Class I and III uniforms wore their usual cap badge on the ribbon.

CPO's 'Caps, Cloth, with Peak'. These were the same caps as used by Class III Ratings but adorned with CPO cap badges. *At left* is the winter cap with pattern 48 gold wire badge, and *at right* a cap with white cover and metal war economy badge.

A pre-war studio portrait of a young CPO Artificer wearing the double-breasted Class I cloth tunic and cap with white summer cover.

A Petty Officer wears the cloth cap for a studio portrait. He wears the ribbon of the 1939–43 star, on fabric backing cloth, on his left breast.

Class III PO's caps. *At left* the Cap, Cloth, Blue with Peak; and *at right* a cap with the Marcelline cover for summer and tropical use.

Class III Rating's caps. *At left* the Cap, Cloth with Peak, and *at right* a Cap, Cloth with Peak with the Marcelline cover for summer and tropical use.

Class III Rating wearing the Cap, Cloth with Peak fitted with the Pattern 52 cap badge.

A Class III Rating wearing the Cloth Cap with Peak and Pattern 52E badge. The overcoat worn here was an optional item for Class III Ratings.

Class I and III Cap Badges

During the Great War military branch POs had been dressed as seamen, with CPOs wearing the fore-and-aft rig and a peaked cap, similar to that worn by Officers. It bore a cap badge consisting of a silver fouled anchor surrounded by a gold wire rope circle and surmounted by a crown on a black fabric backing patch. In 1919 the Admiralty approved a change of uniform for confirmed POs with four years' seniority, who in future were to wear the fore-and-aft rig as worn by CPOs. The old CPO cap badge was adopted by POs, and a new pattern was approved for CPOs, being a silver metal fouled anchor surrounded by a rope circle within a laurel wreath and surmounted by a crown, all in gold wire. The new badges were introduced in the summer of 1920.

Woven red on black cap badge worn by Class III Ratings and Acting POs. *At left* is a pre-war padded Pattern 52 cap badge, and *at right* is the flat type, wartime economy, Pattern 52E badge.

Pattern 49 cap badges as worn by POs. *At left* is the standard pattern and *at right* the metal wartime economy version introduced in 1941.

Pattern 48 cap badges worn by CPOs. *At left* is the standard embroidered pattern, and *at right* the metal wartime economy pattern.

Since the introduction of fore-and-aft rig for Class III Ratings in 1890 they had worn a red woven version of the CPO cap badge. From 1919 the non-military branch Class III POs were authorized the use of the new gold wire PO cap badge when confirmed, while other Class III Ratings continued to wear the previous pattern of red badge.

In 1939, just prior to the outbreak of war, the three standard patterns of Ratings cap badges were the gold wire pattern 48 worn by CPOs, the gold wire pattern 49 worn by POs, and the woven red-thread pattern 52 worn by Class III Ratings. In May 1941 the increased difficulty in obtaining an adequate supply of embroidered CPO and PO badges saw these badges supplemented by metal badges of a similar design. The new metal badges retained the same pattern numbers and issuing prices of the woven pattern, and demand was to be met by the provision of whichever type of badge was available. The metal CPO and PO cap badges had the embroidered gold wire section replaced by gilt metal, with a padded crown mounted on a black fabric backing patch. An economy version of the Class III badge was also produced. The 'Class III flat' (austerity pattern) E52 badge was a simple flat weave badge, whereas the standard pattern 52 had padded and raised anchor and crown devices.

Class I and III Tropical Clothing

AFO 587/39 stated that No. 10 Dress tropical rig for Ratings was in future to be as follows: CPOs and confirmed POs, tropical shirt, white shorts, white stockings and white canvas shoes. Other Ratings not dressed as seamen were to wear tropical shirt, blue stockings and black boots or shoes. The use of white canvas shoes and white stockings by all ranks could be authorized by the Commander-in-Chief in extremely hot weather. The scale of issue was three pairs of shorts and three shirts or tropical singlets, and one pair of white canvas shoes.

An interesting photograph of signallers stationed in Malta *c.*1939. The CPO seated at front right wears the gold-on-black pattern 7c CPO visual signaller 1st Class badge on the right cuff of the No. 7 Dress drill tunic. The leading signallers standing behind wear a most interesting array of cap tallies. *From left to right* they are HMS *Sussex*, *Glorious*, *Cossack*, *Galatea*, *Shropshire*, *Intrepid*, *Barham* and *Warspite*.

A studio portrait of a CPO wearing the white drill tunic.

Class I and III Drill Tunic

The drill uniform worn by Class I and III Ratings as No. 6, 7 and 5 Dress consisted of a single-breasted white drill tunic with four-button front, mandarin collar and an open top pocket fitted at each breast. CPOs and confirmed POs wore gilt fastening buttons. CPOs wore three ⅞in buttons at each cuff; other Class III Ratings wore ⅞in white horn buttons and no cuff buttons. Fly-fronted white drill

The pattern 77c CPO photographer non-substantive badge featuring a bellows camera surmounted by a crown.

Detail of the CPO's pattern white drill tunic worn as No. 6 and 7 Dress. With blue cuff badge, the tunic was worn as No. 5 dress. This is the post-1936 pattern of tunic.

Chief Petty Officer's drill tunic worn by a CPO Photographer 'Snaps'. The tunic has three buttons on each cuff and a gold-on-blue pattern 77c CPO photographer non-substantive badge.

ABOVE: Substantive pattern 1B-rate badge for a PO worn on the left sleeve above a single pattern 42B & C, good badge (denoting three years, good conduct). The badges are in blue on white as worn with No. 6 and No. 7 Dress.

LEFT: PO stoker wearing the pre-1936 pattern Tunic, Drill as No. 6 Dress uniform. This was worn instead of No. 1 Dress in hot climates. The front has a four gilt crown and anchor-buttoned closure. The Uniform was also worn as No. 7 Dress.

ABOVE: Non-substantive crown over propeller pattern 6B badge worn by PO stokers on the right sleeve. Blue on white as worn with No. 6 and No. 7 Dress.

RIGHT: A Petty Officer gunnery Rating wearing the white drill uniform with gilt buttons.

BELOW: A Class III Rating wearing the white high collar No. 7 or No. 6 Dress. The tunic has four buttons of white horn, and the cap is worn with a Marcelline cover and Pattern 52 badge.

trousers were worn with the drill tunic. The tunic cut was improved in September 1936. Among the alterations was the collar being set lower, a second hook was fitted for collar closure, and a throat tab was added. The pockets were raised slightly, and the corners were cut square. The trousers worn with the drill tunic were modified in November 1937, being fitted with belt loops so that they could be worn with belts or braces.

In white uniform CPOs wore non-substantive badges on the right cuff above the centre gilt button. Gold badges were worn with No. 6 and 7 Dress, and blue with No. 5 Dress. Other Class III Ratings wore badges in blue on the upper sleeves.

Singlet, White Cotton

The Singlet, White Cotton, with Spinal Piece (Singlet, Cotton, Tropical; Singlet, Tropical) is described in detail in the Class II section.

Shirt, Tropical (Shirt, Tropical, CPOs)

The Singlet, White Cotton, with Spinal Piece was issued to all Class I, II and III Ratings until October 1932 when a new open-neck shirt with roll collar – the Shirt, Tropical – was introduced for CPOs under AFO 2351. The tropical shirt was a smart, short sleeved, half-front cotton shirt with two open top pockets at the chest. It had a single button fastening closed with a gilt button. It was often called the Shirt, Tropical, CPOs, but by the spring of 1939 the use of the shirt had been extended to include all Class III Ratings. CPOs and confirmed POs wore a 9⁄16in gilt button below the neck opening of the shirt. Other Class III Ratings wore a black horn button.

PO visual signalman No. 10 Dress. This PO wears white shorts with the Shirt, Tropical fitted with a gilt button, and cap with Marcelline cover.

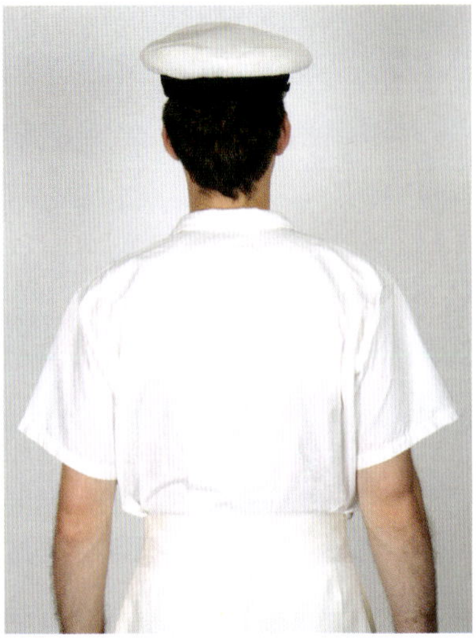

Back of the Shirt, Tropical, showing the spinal piece fitted to protect the wearer from the heat.

The Shirt, Tropical, in wear. The button has been left open, showing a vest worn beneath.

Sun Helmets and old pattern tropical singlets worn by sailors and Marines serving with the Mediterranean Fleet. Class I and III ratings are indistinguishable in this inter-war image, taken prior to the introduction of Sun Helmet badges.

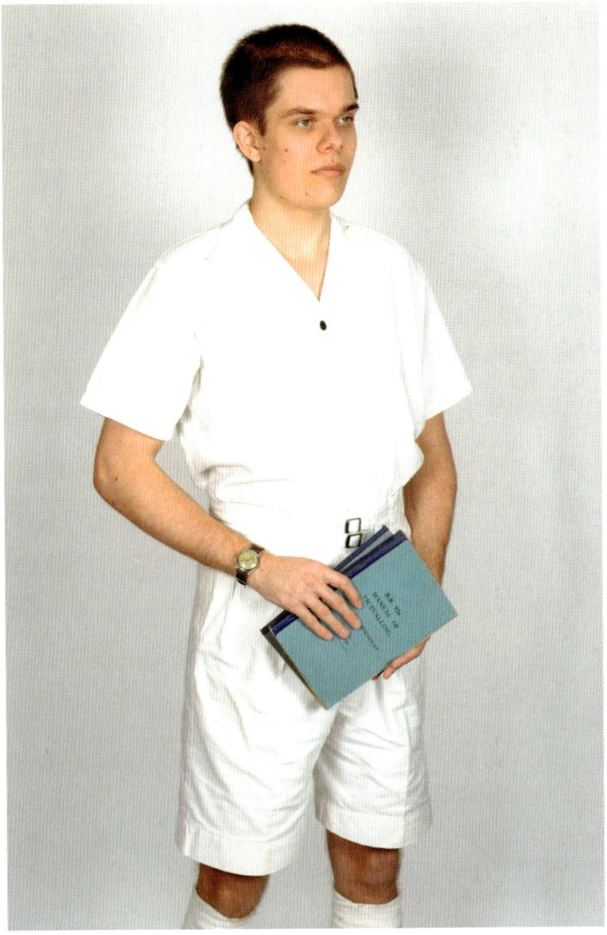

No. 10 Dress worn by a Class III Other Rating. White stockings and canvas shoes, introduced in 1932, could be worn with tropical clothing in lieu of blue stockings and black shoes when authorized.

Class III Other Ratings No. 10 Dress. Class III POs and Leading Rates wore their substantive rate badge on the left sleeve of the Tropical Shirt.

Class II Uniform

Class II uniform 'square rig' was worn by all POs and men dressed as seamen. POs dressed in Class II uniform were acting in the Rating – men with less than twelve months in the Rating. Upon confirmation as PO, after twelve months acting, they would exchange their square rig Class II uniform for the Class III fore-and-aft rig.

Class II Dress No. 1: For inspections, musters, ceremonial occasions, Sundays in harbour and when proceeding on leave, the No. 1 Dress was worn by men dressed as seamen. It consisted of a serge suit with gold badges and medals.

Class II Dress No. 2: The No. 2 Dress was worn for Sundays at sea and by signal Ratings on duty on Sundays when in harbour. It was optional clothing for weekday wear when on leave. All duty men, boats' crews and signallers were to wear the No. 2 Dress when the ship's company were in No. 3 Dress. For men dressed as seamen, No. 2 Dress consisted of a serge suit with red badges and medal ribbons.

Class II Dress No. 3: For normal working-day wear the No. 3 Dress was worn by men dressed as seamen. It consisted of a serge suit with red badges. Medal ribbons were optional with No. 3 Dress.

Class II Dress No. 4: For wet weather and for normal evening wear (night clothing) the No. 4 Dress was worn. Any serge suit with red badges could be worn as No. 4 Dress. As with No. 3 Dress, medal ribbons were optional.

Class II Dress No. 5: For hot weather the No. 5 Dress replaced the No. 3 Dress as normal working clothing. It could also be worn when No. 3 clothing was not considered suitable. It consisted of the white drill (working) jumper and white drill trousers. Medal ribbons were not to be worn. White drill clothing was the standard working dress of boys in harbour training establishments.

Class II Dress No. 6: At the discretion of the senior Officer, No. 1 Dress could be replaced by the No. 6 Dress on tropical stations. For No. 6 Dress the white drill jumper with collar and white drill trousers were worn with medals.

Class II Dress No. 7: In warmer climates No. 2 Dress could be replaced by the No. 7 Dress, at the discretion of the senior Officer. In No. 7 Dress the white drill jumper with collar and white drill trousers were worn with medal ribbons. Prior to 1933, No. 7 Dress had been as for No. 2 Dress, but with drill trousers or as No. 6 without medals, as ordered.

Class II Dress No. 8: In June of 1939 there was no No. 8 Dress for men dressed as seamen. No. 8 Dress had previously been as for No. 3 Dress, but with duck trousers, or as No. 6 without medals, as ordered. It was discontinued in 1933.

Class II Dress No. 9: For particularly dirty work such as coaling ship, painting or refitting, and when other classes of clothing may be spoiled, men dressed as seamen would wear the No. 9 Dress. This consisted of the blue overall suit (Combination Overalls).

Class II Dress No. 10: In tropical stations on which the use of tropical clothing was allowed, the wearing of No. 10 Dress was authorized at the discretion of the Commander-in-Chief. For men dressed as seamen No. 10 Dress comprised the tropical singlet and white shorts worn with the waistbelt and white cap or Sun Helmet as ordered.

Class II Dress No. 3A: As for Class I.

Class II Dress No. 8: As for Class I.

Uniform Kit List Class II

The following lists are for Petty Officers and men dressed as seamen, June 1939.

Compulsory Kit

Three jumpers, serge
Three pairs trousers, serge
Two jumpers, drill
Three jumpers, duck (seamen)
Two jumpers, duck (stokers, signallers)
Two pairs trousers, drill
Three pairs trousers, duck (seamen)
Two pairs trousers, duck (stokers, signallers)
One blue overall suit (seamen)
Two blue overall suits (stokers, signallers)
Three collars, blue jean
Two caps (Ratings on foreign service or on home service from 1 May through to 30 September)
Three caps (Ratings on home service from 1 October through to 30 April)
One box, cap
Two cap ribbons
Two scarves, black silk
Two pairs socks or stockings
Two pairs boots, half, or one pair boots, half, and one pair of shoes
One bed
One blanket
Two bed covers
Three cotton flannels
Three singlets for wear under flannels
One jersey
One knife (compulsory for seamen branch but optional for other Class II Ratings)
Two lanyards, knife
One bag, soap
One comb, hair
One brush, hair
One brush, tooth
One brush, clothes
One brush, boot, hard
One brush, boot, polishing
One type, name
One ditty box or attaché case
Two pairs drawers
Two towels
One waistbelt, blue
One oilskin coat
Volume 1 Seamanship Manual (ordinary seamen and boys only)
Stokers Manual (stokers only)
Signal card (signal Ratings only)

Optional Kit

One overcoat
One waterproof coat
One pair oilskin leggings
One sou'wester
One jacket, canvas (stoker Ratings only)
One pair trouser, Fearnought (compulsory for stoker Ratings on coal-fired ships)
One pair shoes, black leather
One pair shoes, brown canvas, or slippers, black, leather
One pair shoes, white canvas
Two pairs socks, white
Two suits, pyjamas
One razor or pair of scissors
One pair gloves, woollen
Six handkerchiefs, white cotton
One brush, shaving
One brush, boot, blacking
One housewife
Two belts, cholera
One pair gaiters, black leather
One comforter
One cap cover, Marcelline
One scarf, plain white
One pillow
Two pillow covers
Two sheets, white
One suitcase
One waistbelt, white
One pair rubber seaboots
One cover (washable) for gas mask haversack

Tropical clothing (when ordered)

No. 10 Dress: men dressed as seamen
Tropical singlet
White shorts
Blue stockings
Black boots or shoes

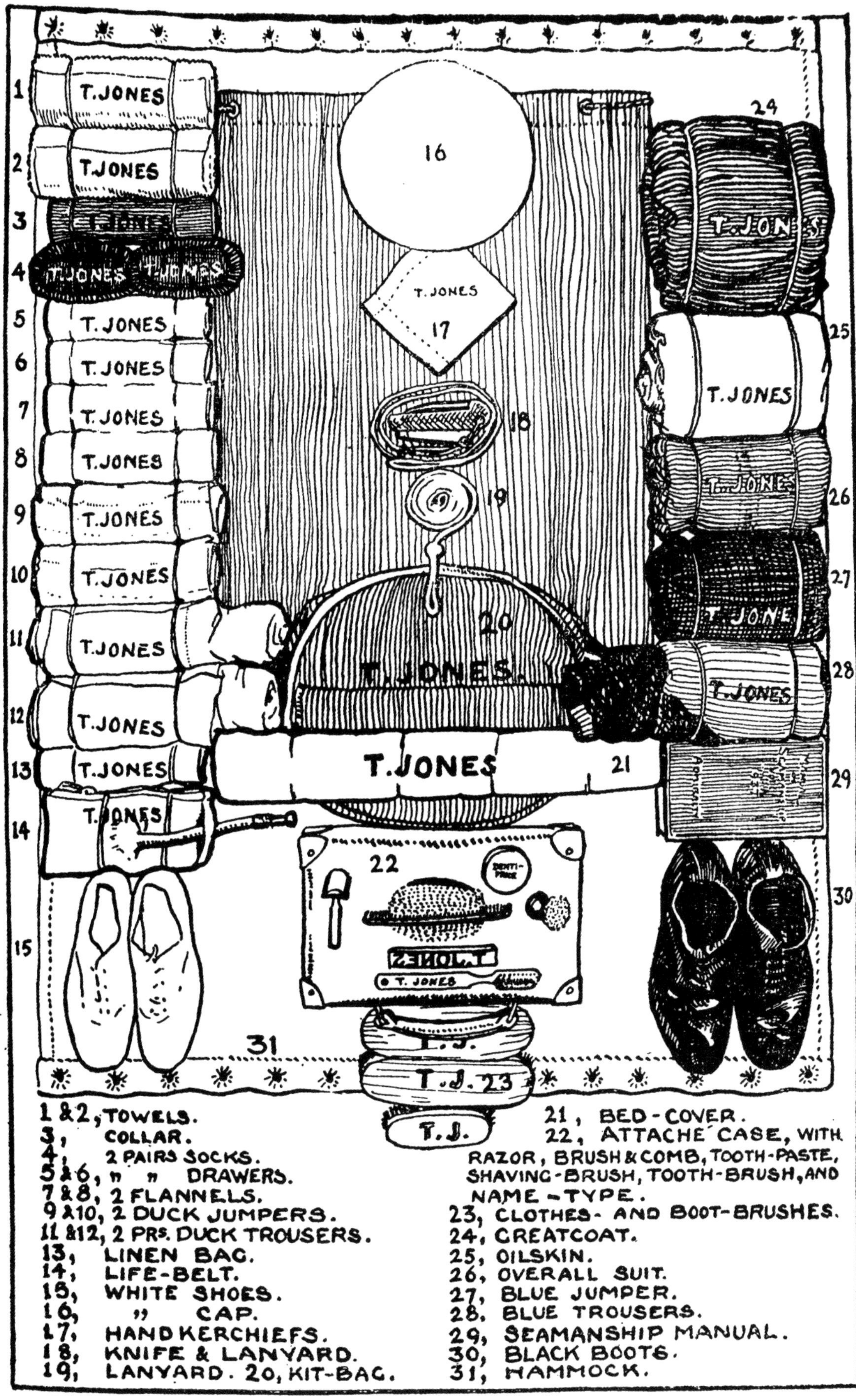

A Class II seaman's kit laid out for muster as shown in the Seaman's Pocket Book B.R.827. The clothing is representative of that issued to a hostilities-only Rating wearing a serge suit, collar and silk, flannel, drawers, boots and blue cap at the time of muster. All items of clothing have been rolled and tied so that the Rating's name is readily visible.

Class II Uniform Items

The contents of AFO 1475 (alterations to the holdings of Ratings uniform and kit in the event of mobilization) were actioned in October 1939 when AFO 3025 confirmed the modified compulsory kit requirements of Class II Ratings for the duration of the war. The wearing of No. 1 Dress and gold badges was suspended except as optional for wear while on leave. In addition, No. 6 and No. 7 Dress (white drill jumper with integral blue jean collar and blue jean cuff and hem trim) were also suspended, except for optional wear on leave. In place of the drill clothing the white duck working jumper and trousers were to be worn with a blue jean collar as required. In practice the drill jumper continued to be used by men who possessed or could obtain them.

The quantity of compulsory clothing held was also reduced. Holdings of serge and duck suits were reduced from three to two, with a single serge suit and the two drill suits moved from the compulsory to the optional kit list.

Jumpers and Trousers

Jumper, Serge, with Cuffs
Jumper, Serge, Working (Jumper, Serge, without Cuffs)
Jumper, Serge
Trousers, Serge

Prior to 1932 two patterns of serge jumper were issued to Class II Ratings: the Jumper, Serge, with Cuffs, and the Jumper, Serge, Working. A single Jumper, Serge, with Cuffs was compulsory kit for each Class II Rating: it had cuffs with two buttons for fastening and was worn with gold badges as No. 1 Dress. Two Jumpers, Serge, Working, were required to be held. Worn with red badges, they were used for No. 2, No. 3 and No. 4 Dress. The Jumper, Serge, Working was similar to the Jumper, Serge, with Cuffs, but had plain sleeves with no cuff. The sleeves were 1in shorter than those of the jumper with cuffs. Old Jumpers with cuffs that were no longer suitable for wear as No. 1 Dress could have the cuffs removed for use with No. 2, No. 3 and No. 4 Dress. The length of the V front had been 11in for all sizes, but in 1926 this was changed to 11in to 13in depending on height.

In 1932, as a result of the report of the Uniform Committee 1931, several modifications were made to the Class II jumper and trousers. The V front opening of all jumpers was modified to extend to 1in below the breastbone, providing an opening of 12 to 15in in depth, and making the jumper slightly easier to pull on. Prior to 1926 the depth of the V had been 11in for all sizes, proving too short for taller men and too long for shorter individuals. Between 1926 and 1932 the length was set at 11 to 13in, depending on the height of the wearer. The bottom edge of the jumper was lengthened slightly to be in line with the crotch, or not more than 2in above it. The Jumpers, Serge, Working were removed from compulsory kit, and in future all serge jumpers were to be fitted with cuffs and tailored to the body to improve fit at the waist and hips and to provide a smarter form than the near-parallel sides of the old-style jumper. The drill and duck jumpers were in future both to be looser fitting than the serge jumper, and were to remain without cuffs.

A review of uniforms in the Mediterranean Fleet had resulted overwhelmingly in a request for the removal of the internal pocket from the jumper. However, other units had stated that it was useful. The Admiralty therefore compromised and decided that the fitting of a pocket on the inside of the jumper was optional for trade-tailored jumpers, but that those from service sources would retain the pocket.

Class II 'bell-bottom' trousers were also modified as a result of the clothing review. In 1932 the width was set at 12 to 13in at the knee, and 12 to 14in at the ankle, depending upon the build of the individual. Prior to 1932 the trousers had been parallel-sided, 12in wide at both the knee and bottom edge across all sizes. In addition the serge trousers' pockets and waist lining of blue jean was henceforth to be replaced by blue and white striped shirting, except for the overlap, which was to remain blue jean. The wide bell-bottom of the Class II trousers allowed them to be rolled to the knee for scrubbing the decks (a daily routine), and allowed easier removal if the wearer fell overboard.

Both the Class II jumper and trousers were normally stored inside out and tightly rolled. When unrolled this gave an inverted crease on the jumper sleeves and at each side of the trouser leg. It was also common practice for the trousers to be folded in a concertina fashion to the width of the S.34A Pay Book, resulting in seven or five creases dependent upon height.

In 1937 a batch of several 'Jumpers, Serge' were fitted with a zipper fastening to the front. The trial jumpers came about following numerous requests for a garment easier to put on. However, the Admiralty made it clear that the difficulty in putting on and removing the jumper was due to the fashion for the jumper to be tailored tighter and tighter over the years, and added that it was unlikely that a zip would stand up to the treatment to which the jumper was subjected, and might be too costly to approve. Ultimately the zip fastener was not approved, and the design fell by the wayside until 1956 when it was resurrected as the Class II 'coat-type' serge jumper. The trousers' fall front was also discarded in favour of a fly front the same year.

In May 1941 the serge and duck jumpers were modified by the introduction of an extra pocket on the right inside breast to supplement the single pocket previously only fitted on the left breast. The additional pocket was expressly for the carriage of the S.43A *Sailor's Pay Book and Identity Book* and was required to be large enough to fit the 5 by 4in book. Jumpers in possession of men were ideally to be modified by the addition of a new pocket, and privately purchased trade jumpers were to be fitted with the pocket. A button was later added to the specification so that the left pocket was open top and the right had a button closure.

By the spring of 1942 the collars of serge jumpers were being produced in a single thickness of fabric, with a dungaree edge binding, rather than the double thickness previously used. This greatly reduced the fabric required in manufacture.

AFO 2550 of June 1943 introduced further changes to the Class II uniform jumper and trousers. The cut of the sleeves was to be improved, and as a wartime economy measure button cuffs were to be omitted from all future production of serge jumpers. The two inside breast pockets were both henceforth to be fitted with a closure button. The method of attaching the serge collar to the jumper was also to be strengthened, and the blue and white striped 'shirting' lining of the Class II trousers was to be replaced by linings of plain unbleached calico. The paper size labels that were previously fully stitched to the rear lower skirt of the jumper and the lower leg of the trousers were merely tacked on to the modified austerity garments with a large 'X' of cotton, further reducing cost and labour time. As well as clothing purchased from Slops, these modifications were also to be embodied on all trade items purchased from private tailors.

In January 1944 AFO 154 reiterated the continuing problem of deviation from standard sealed pattern uniforms that often occurred when clothing was purchased from commercial trade sources ashore. Problems encountered included jumpers cut with a distinct 'U'-shaped front, the front opening cut too low, and trousers cut unnecessarily full in the leg. Wool had been in short supply since 1942, and such wasteful deviations in pattern could result in the Ratings concerned being disciplined for non-observance of dress regulations.

The spring of 1945 saw a slight relaxation of the austerity measures applied to the production of Class II serge jumpers. AFO 1222 of March 1945 reintroduced the buttoning cuffs to the serge jumper, and all-new production jumpers were henceforth to be manufactured with cuffs. Typically, stocks of the austerity pattern with plain sleeve had to be used up before the new button cuff pattern could be issued.

It is worthy of note that the paper size labels applied to RN garments can be misleading when dating production. H. Lottery & Co. Ltd were one of the primary RN clothing contractors during World War II, and their labels may be found on a multitude of garments. It would appear that the company massively overstocked on these labels during the early part of the war, as 1941-dated labels regularly appear on post-1943 production austerity clothing. Some garments produced as late as 1945 still had 1941-dated labels. The labels can sometimes be found overwritten by hand with the actual year of manufacture, but more often than not, no amendment was made. Such labelling has caused confusion among collectors and historians.

Evolution of the Seaman's 'Jumper, Serge'

Jumper, Serge, with Cuffs. 1926–32. Button cuffs. Single inner pocket, open top. 11–13in V front.

Jumper, Serge, Working. 1926–32. No cuffs. Single inner pocket, open top. 11–13in V front.

Jumper, Serge. 1932–41. Button cuffs. Single inner pocket, open top. 12–15in V front.

Jumper, Serge, Zip Front. 1937 trials. Button cuffs. Single inner pocket, open top. 12–15in V with zip front.

Jumper, Serge. 1941–42. Button cuffs. Twin inner pocket. 12–15in V front.

Jumper, Serge. 1942–43. Button cuffs. Twin inner pocket, one open, one button top. 12–15in V front. Single-thickness collar with bound edge.

Jumper, Serge. 1943–45. No cuffs. Twin inner pocket, both button top. 12–15in V front. Single-thickness collar with reinforced bound edge.

Jumper, Serge. 1945. Button cuffs. Twin inner pocket, both button top. 12–15in V front. Single-thickness collar with reinforced bound edge.

LEFT: Unchanged since before the Great War, the pre-1932 'Jumper, Serge, with Cuffs' with characteristic shallow V front. The 'Scarf, Black Silk' is bloused, with the free end secured by tapes and the bight tucked inside the jumper's V, a common but unauthorized practice. Rigged for wear by a Quartermaster, the Bosun's call is tucked into the jumper front.

The rear of the waistband of early 1930s production 'Trousers, Serge' showing the adjustment lace used in all Class II serge, duck and drill trousers.

Seaman Gunner badge Pattern 115 of the pre-1938 organizational system.

The internal pocket of the pre-1932 'Jumper, Serge, with Cuffs', showing the method of hanging the pocket, almost parallel to the V. The end of the V front is just below the level of the armpit on jumpers produced prior to 1932. This style of pocket shape and fitting remained in use until 1941.

German Reichsmarine sailors from the cruisers *Konigsberg* and *Leipzig* fraternize with British sailors during a visit to Portsmouth in July 1934. The British seamen wear the Ratings No. 8 Dress, a mix of blue jumper and white duck trousers which had been removed from Dress Regulations in 1933.

The pre-1932 Jumper, Serge, Working. The working jumper had plain cuffs with sleeves that finished 1in above the wrist bone, markedly shorter than on the Jumper, Serge, with Cuffs.

A noticeboard photograph of gunners at HMS *Excellent* showing the new Class II Ratings' Uniforms. *At left* is the old 'Jumper, Serge, with Cuffs', and *at right* the new pattern 'Jumper, Serge' introduced in 1932. The new jumper was longer, better tailored, and had a deeper V opening to the front. The new cap sat better than the preceding pattern and had considerably improved stiffening.

PO torpedoman. This acting PO, of less than one year in the rating, wears the gold badge 'Jumper, Serge' as No. 1 uniform. He carries the wool overcoat and attaché case.

RIGHT: June 1939, RNAS Lee-on-Solent. A trainee air observer swings the prop prior to the engine being started. The Class II serge jumper and trousers were a somewhat impractical flying dress, but the ensemble was the standard working dress for Ratings.

Above is the interior of a commercial Class II 'Jumper, Serge' showing the popular U-shaped front. The deviation from service pattern was not authorized. Below is a single pocket Admiralty production jumper fitted with a second pocket following its introduction in May 1941.

MTB Ratings wearing Class II uniform. *At left* is a Seaman Torpedoman wearing the good shooting badge on his lower right cuff and H.M.M.T.B. cap ribbon. *Centre* is a Wireless Telegraphist 3rd class with H.M.T.B. cap ribbon. *At right* the Rating wears a post-June 1943 jumper with plain cuffs and extravagantly long tapes.

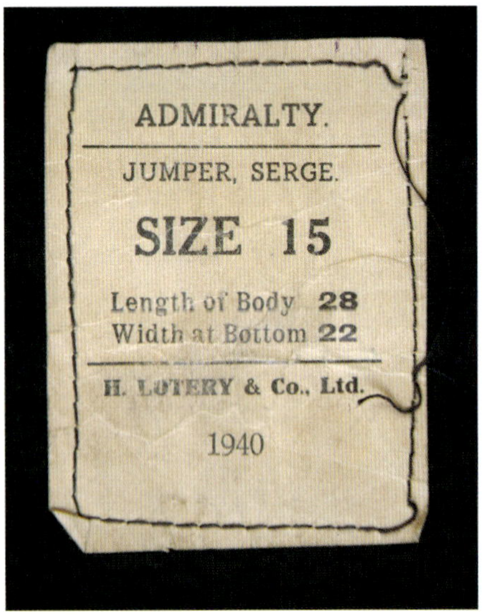

FAR LEFT: The paper label from a Class II Jumper, Serge. Dated 1940, it was common wartime practice for old labels to be used up rather than wasted; the size, and sometimes the date, was altered by hand to suit the garment.

LEFT: Typical paper label fitted to the inner lower leg of service issue 'Trousers, Serge'.

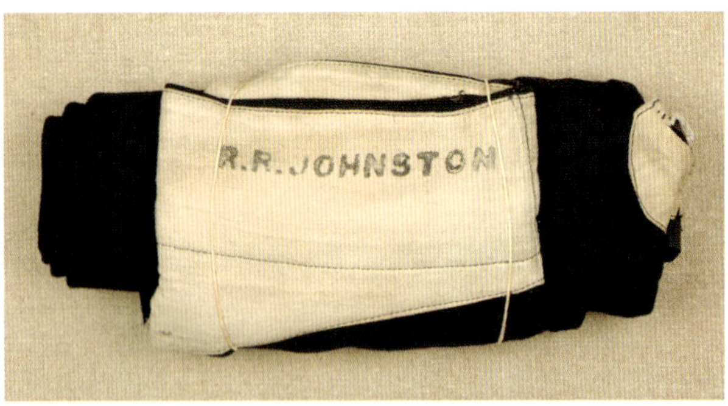

Post June 1943-manufactured Trousers, Serge, with a lining of plain unbleached calico. The trousers are rolled and tied for kit muster with the name clearly visible.

Acting PO wearing the Class II rating's No. 1 Dress Jumper, Serge, with blue jean collar and wool jersey worn beneath. The silk scarf and lanyard are secured at the V of the jumper in the prescribed manner, with the cloth tape tied in a bow.

LEFT: HM the King reviews divisions during a visit to HMS *Glasgow* in March 1943. The Class II Ratings are 'off caps' at attention. The nearest Rating wears a No. 2 Dress red badge suit, while the senior Rating to the rear retains his gold badge double-breasted suit.

The waistband of a pair of post-1932 Class II serge trousers showing the fall front and pocket on the right side. The lining was of blue and white striped shirting with an overlap of blue jean. Brace buttons were fitted, but rarely used, as it was customary for trousers to be tailored close fitting.

A 'three badgeman' photographed on his wedding day. On the Class II Rating's wedding day, black uniform tape was always replaced by a white silk tape, and often accompanied by the white silk scarf.

The pattern 123 Leading Torpedoman non-substantive badge in gold wire as worn on his right sleeve by the acting Petty Officer shown on page 88 centre right.

Detail of the pattern 134A Seaman Torpedoman badge as worn on the upper right sleeve by the Torpedoman shown on page 90.

The top image shows the pocket arrangement of a post May 1941 Jumper, Serge, showing the pockets fitted to the inner breast: one open top and one with button closure. *Below* is the inside of a war-economy Class II 'Jumper, Serge' showing the twin pockets with buttons fitted to all jumpers from 1942.

Seaman Torpedoman wearing No. 3 Dress with white cap and cotton flannel. No lanyard was worn in No. 3 Dress. He wears the war economy variant of the serge jumper with plain rather than button cuffs.

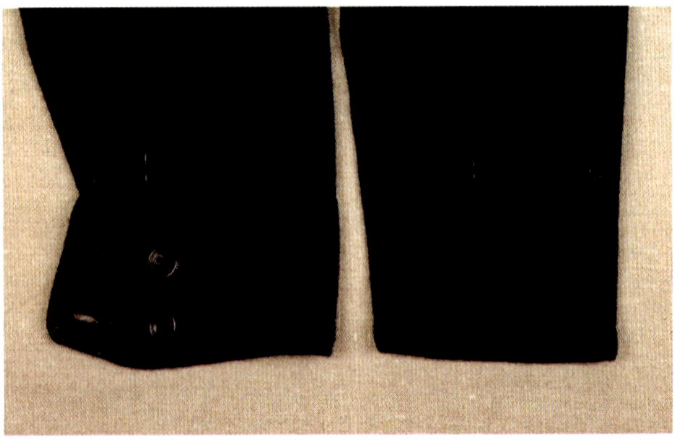

Cuff details of the Class II Jumper, Serge, showing *at left* a pre-June 1943 jumper with twin cuff buttons and blue jean lining, and *at right* a post-June 1943 manufacture, war-economy jumper with plain cuff.

Integral collars of wartime Jumpers, Serge. Above is the style manufactured between 1942 and 1943 showing the single thickness of fabric with dungaree-edge binding. The binding was reinforced with an additional V-section of dungaree at each shoulder in June 1943, as shown below.

Collar, Blue Jean

The seaman's collar in one form or another dated back to the 1830s. In 1845 the Admiralty issued blue jean fabric and white Dutch tape so that men could add blue jean collars, cuffs and facings to their duck frocks. No pattern details were provided so a variety of styles were made, particularly noticeable in the number of rows of Dutch tape. This situation wasn't rectified until 1856 when the Rear Admiral Committee recommended that the white duck frock should have two rows of white tape on the blue jean cuffs and collars. However, canvassing of the Commanders-in-Chief of Devonport and Portsmouth and the ships' captains under their command resulted in a majority verdict that three rows should be standard. The collar was henceforth to have three rows of $^3/_{16}$in white tape at the edge, with a $^1/_8$in gap between the rows. While the original collar had been a part of the duck uniform, it was soon provided for serge uniforms, first as an integral collar and then as a separate item.

A rather odd contraption, the collar was placed over the flannel before donning the jumper. The rear of the collar lay on the shoulders, while a central section draped down the spine. Two extension pieces ran to the front over the shoulder, through a half twist at the front and with two tapes that were then passed through loops on the spinal section before being fastened with a bow at the front. The two extension pieces were connected at the front by a small strip of fabric. To improve the appearance of the collar front this strip was normally cut allowing the front sections of

Front and rear views of a post-1941 'Collar, Blue Jean' with composite edge tape. The small strap joining the two front sections of the collar was usually cut, allowing the lower ends to be pulled to the side, thus allowing for a sharper cut-away into the front of the serge jumper.

the collar to 'cut away' inside the jumper and much higher than would otherwise be possible.

The front face of the collar was made of blue jean with a backing fabric of blue and white striped shirting. The edge of the collar was bordered by three rows of white tape, sewn by hand. The 1926 and 1932 modifications to the Class II jumper both resulted in minor changes to the length of the blue jean collar so that it was compatible for wear with the cut of the modified jumpers.

In October 1941 AFO 4541 announced that in order to facilitate production, seamen's blue jean collars would in future have the white edge tape sewn on by machine rather than by hand as had previously been required, and instead of the three single rows of white tape a new, single section, composite tape would be used. The composite tape consisted of three white bars alternating with two blue bars. Stocks of the old pattern collars had to be exhausted before the new pattern was issued.

An MTB crewman wearing No. 3 Dress with a well scrubbed and faded collar. The white duck cap bears a commercial pattern, H.M.T.B. printed ribbon.

The three rows of white edging of seamen's collars were stitched on by hand, as shown by the upper collar. In 1941 a new composite tape was introduced incorporating three white bars with two blue bars. The composite tape was machine stitched to the collar edge, although many were hand stitched.

The label from a wartime production collar showing the manufacturer and date of 1943.

The blue jean collar as issued was a deep royal blue colour, but it was fashionable for it to be scrubbed and boiled to a well weathered and salty light blue finish. Collars were produced in a ready-to-wear lighter shade of blue by trade outfitters, who also supplied collars made of lightweight cotton.

Handkerchief, Black Silk (Scarve, Silk)

Prior to August 1934 the Handkerchief, Black Silk was square, but from that year it was manufactured as an oblong of 50 × 12in. The name was changed to 'Scarve, Silk', although from 1943 it was manufactured from rayon. The silk was not required to be worn with uniform when coaling or refitting.

Regulations required that the silk was tied at the back of the collar with a reef knot, or stitched to length. The bight (loop) of the free end was to be three fingers in length hanging below the bow of the tape; it was not to be tucked inside the V of the jumper, though it was common for it to be worn in such a manner.

Jerseys, Woollen, Class II

A woollen jersey was provided for winter wear with the serge jumper. Often called the sea or seaman's jersey, it was a tightly knitted, Guernsey-style garment worn with Nos 1, 2, 3 and 4 Dress by all Class II seamen. Once the wearing of the jersey had commenced with the onset of cold weather it was worn until the beginning of April.

Dickey, Woollen

A 'dickey' version was made by onboard tailors, where one wool jersey would make two dickeys with a cotton back. On warmer winter days and below deck the wool dickey was cooler than a wool jersey when worn with the serge jumper.

Although less common than the cotton dickey, the wool jersey was sometimes cut down to form a dickey with open tie-closure sides.

The Class II wool jersey worn by an MTB crewman as outer clothing. At action stations and closed up at his Oerlikon, anti-flash and steel helmet are worn.

A young Rating wearing No. 2 Dress with the Class II wool jersey and serge jumper for winter wear. Of interest is the HM DESTROYER cap ribbon worn on the blue cloth cap.

Class II wool jersey. The jersey was worn beneath the serge jumper, for winter wear. Against regulations it was often worn as outerwear or night clothing at sea.

The Class II wool jersey worn by small ship Ratings engaged in supplementing their rations with fresh fish.

Flannel, Wool (Flannel, Vest or Flannel, Service)

The wool flannel was a shirt-type garment with a broad square neck opening not to exceed 23in and trimmed with ½in blue jean fabric. In 1932 modifications to the Class II serge jumper resulted in minor changes to the cut of the flannel so that it was more comfortable when worn with the new pattern jumper. The poor cut of the old-style service issue jumper and the excessive length of the flannel (as issued it came almost to the wearer's knees) led to Ratings looking dishevelled, often with a distinct bulge at the waist. To help remedy this, the flannel was re-tailored to have less length in the body and shorter sleeves. The wool flannel was worn with Class II dress serge, drill or duck suits.

Flannel, Cotton

When used with the white drill suit in tropical climates the wool flannel proved exceptionally hot and often caused prickly heat. To overcome this problem, in March 1931 a cotton version of the flannel was authorized for local procurement on foreign stations, at the individual's expense.

Boys undertake gun drill. The boys wear a training rig of duck trousers with waistbelts, wool flannels and web anklets with ankle boots. The pre-1932 pattern of flannels have sleeves coming to below the elbow. Post-1932 flannels had shorter sleeves that finished above the elbow.

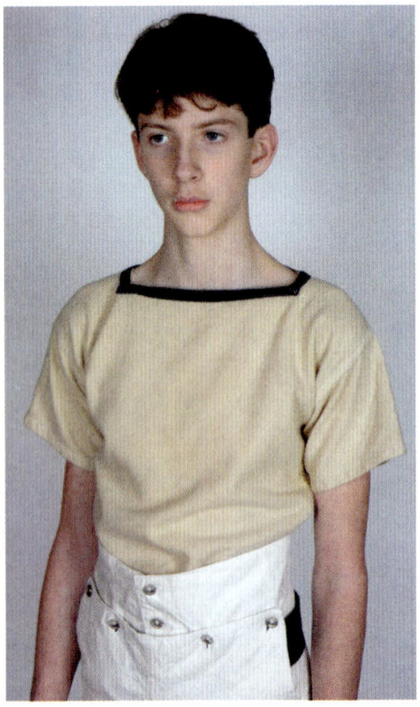

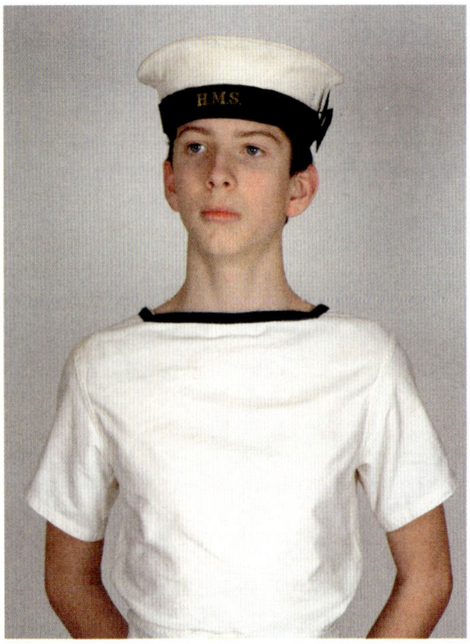

FAR LEFT: A post-1932 pattern, short-sleeve Wool Flannel worn by a Boy 2nd Class. The Wool Flannel was discontinued for general service wear in 1938 but remained as an item of dress for trainees, being renamed the Training Flannel. The blue waistbelt is also worn, in the approved manner, beneath the front flap of the duck trousers.

LEFT: A Boy 1st Class wearing the Cotton Flannel. During the war boys approaching draft to sea service from harbour training establishments were required to replace the Training Flannel with the standard Cotton Flannel, as new kit was required or prior to joining a sea-going ship. Training Flannels could be worn with Working Dress at sea until worn out.

The following year a standard pattern cotton flannel was introduced. The 'Flannel, Cotton', made of bleached cotton twill, was similar to the wool flannel but was for use only on tropical stations (other than the Mediterranean) for wear with the white drill uniform. It was not to be used as tropical working dress, for which the 'Singlet, Cotton, Tropical' was used (see below). Flannels issued on the Mediterranean station were made of cotton drill until 1935, when they were replaced by the standard cotton twill flannel.

The Admiralty received frequent requests for the removal of the wool flannel in favour of the cotton flannel. It was finally agreed that this was desirable, and in 1938 AFO 569 replaced the wool flannel with the cotton flannel. The cotton flannel was similar in design to the wool flannel, but was to be used in conjunction with a cellular cotton singlet (vest) to be worn beneath for warmth. Men in possession of the wool flannel were authorized to continue to use them until they were worn out, but no free issue of cotton flannels was made to men already serving. The compulsory Class II kit requirement was modified to show three cotton flannels and three cotton singlets.

Following the introduction of the cotton flannel for men, boys in harbour training establishments continued to be issued the wool flannel as a part of their compulsory kit, being renamed the 'Flannel, Training'.

Trousers and Jumper, Duck, Working (Uniform, White, Working)

The duck jumper and trousers were made of a heavy and hard-wearing duck fabric. They were worn as No. 5 Dress for men dressed as seamen.

In 1936 it was requested that the white drill uniform were provided with a separate collar, that it should be of a lighter fabric, and that old and worn suits should be altered for use as No. 5 Dress. The Admiralty responded by stating that a separate collar for the drill uniform was under consideration, but it was felt that it might prove to be too hot in use. A lighter fabric was not approved as it would be insufficiently hardwearing. With regard to the use of old drill uniform as No. 5 Dress, it was stated that the matter had to be deferred pending the conclusion of the question of the separate collar for the drill uniform, and also trials of a new uniform that might replace the duck working uniform. Boat crews wearing No. 7 Dress had found that the rigours of boatwork caused undue wear on the cotton drill uniform. At the discretion of the senior Officer, the duck uniform could be authorized for use with a separate blue-jean collar by boat crews. However, in 1939 the discontinued use of the No. 6 and No. 7 Dress resulted in the duck working uniform and blue jean collar being authorized in lieu of No. 6 and No. 7 Dress for the duration of hostilities.

Duck to Drill

In 1943 shortages of duck fabric on overseas stations led to the use of drill being approved for manufacturing No. 5 suits if duck could not be obtained. July 1944 saw the official substitution of drill for duck for all working dress jumpers and trousers for Class II Ratings. Other than for a

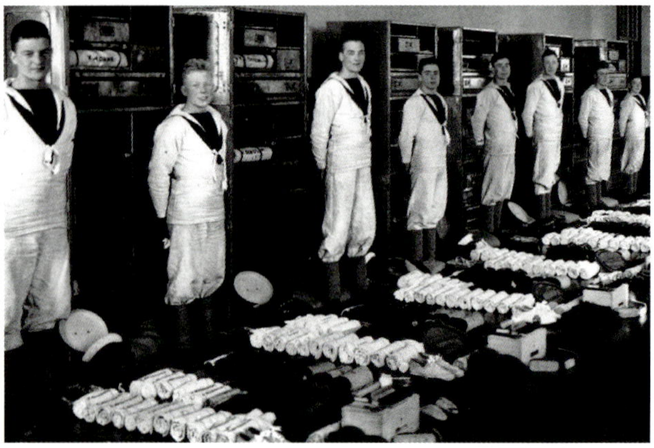

Boys dressed in the No. 5 Dress muster their kit at HMS *St Vincent.* The much despised kit musters were given periodically to all Ratings, with additional musters often being awarded as a punishment.

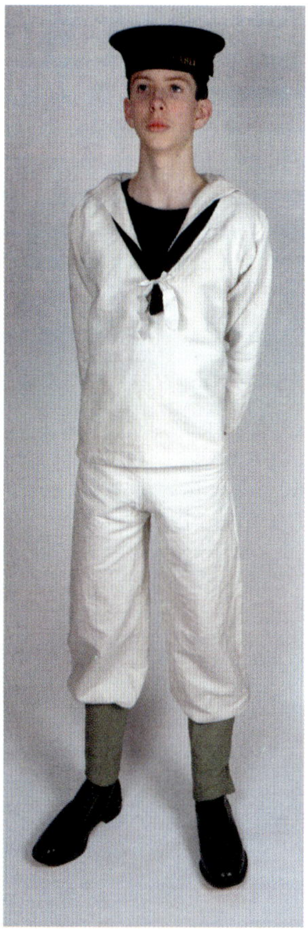

The Duck Working Uniform worn for arms' drill. Despite the evolutions undertaken, the Duck Working Dress was expected to be kept clean at all times.

The White Duck Working Uniform jumper and trousers worn by a boy under training. The uniform is worn with web anklets and blue cap. Typical of trainees, the jumper tapes are pressed in a concertina fashion.

couple of minor modifications, the new drill working uniform conformed to the style of the duck uniform. The drill jumpers and trousers had narrower hems, and the jumper's internal pockets had button closures. The new jumpers were officially titled the 'Drill Jumper No. 5'. The drill jumpers and trousers were only to be issued as the stocks of duck jumpers and trousers became exhausted. Class II compulsory uniform regulations were modified from 'Two jumpers duck, and two pairs trousers duck' to 'Two jumpers duck or drill, and two pairs trousers duck or drill'. In 1946 No. 5 Dress was abolished and stocks of Drill Jumpers No. 5 were converted into No. 6 Dress by covering the existing plain collar with blue jean and composite tape. Blue trim was not added to the cuff or hem on converted jumpers.

Jumper and Trousers, Drill, Uniform

In August 1930 AFO 2070 announced the substitution of duck for drill in the manufacture of No. 6 Dress for men dressed as seamen. The jumper had been called the 'Jumper, Duck, with collar', and it was worn with the same trousers as were used with duck working dress. But after October 1930 new production was to be made from cotton drill. To allow the wearing out of the duck uniform, Ratings were allowed to continue to use the duck No. 6 uniforms until the end of September 1931, after which time only cotton drill was to be permitted. This wearing-out period was later extended to September 1932, and the replacement cost was born by the Ratings from their KUA.

In 1937 it was proposed that the white drill uniform be provided with the standard blue jean collar as a separate item, and the integral blue collar be replaced by a plain white collar. Such a uniform had previously been locally approved for wear on the New Zealand station. The integral blue collar had often been the cause of staining of the white drill jumper, and having a separate collar would alleviate this problem. It was also felt that the jumper with plain white collar could be worn as an off-duty attire on hot stations. The jumper was to retain the blue trim at waist and cuffs, although it was of some concern that this would fade at a different rate than the separate collar, leading to a mismatch in shade.

It was noted that the use of the separate blue jean collar over an integral white collar might cause some discomfort in hot weather, as two collars would effectively be covering the upper back and shoulders. It was recommended that suits meeting the specification should be made up for trials on the Mediterranean, China and East Indies stations. However, the proposed uniform with separate collar was not adopted for Fleet-wide issue. Drill jumpers were available from commercial tailors without integral collars. Although contrary to dress regulations they were particularly popular on the Mediterranean station.

The white drill jumper was of the pull-over-the-head pattern with an integral blue jean collar, having a border of three hand-stitched rows of 3⁄16in white tape. The lower edge of the sleeves and the bottom of the jumper were bound in ½in blue jean. The sleeves of the jumper were worn intentionally short, coming to 1in above the wrist bone. The trousers had a fall (drop) front secured by rustproof white metal buttons. There was a single pocket at the front right side, and a four-hole adjustment lacing in the rear waist. The drill jumper and trousers were worn as the No. 6 and No. 7 Dress for Class II men dressed as seamen.

In October 1939 the issue of drill uniform was suspended for the duration of the war. Ratings already in possession of the uniform could retain it for wear while on leave, but no new uniforms were to be issued, nor were they allowed to be purchased from tailors. If a white uniform was required the No. 5 duck uniform was to be used with separate blue jean collar. When production of the drill jumper eventually re-commenced after the war, the new jumpers featured two button closure pockets fitted internally and a collar edging of composite tape.

Men drafted to ships and establishments abroad were normally to be supplied with all of their tropical clothing before leaving England. However, for ships refitting or commissioning in the USA or Canada, tropical clothing was to be issued once onboard from stocks manufactured and obtained in those countries. To a lesser extent Australian-manufactured clothing was supplied to British ships undergoing repair or refit in Australia.

Both drill and duck jumpers were fitted with twin grommets at both hem seams. The grommets allowed clothes stops to be rove through for hanging the jumpers out to dry.

The Jumper, Duck, with Collar, had remained unchanged from the late nineteenth century until its final demise in 1932 when it was replaced by a lighter and more comfortable jumper in white drill. Twin eyelets at each waist allowed 'clothes' stops' to be fitted when drying the jumper after washing.

The inside of the Jumper, Duck, with Collar, showing the fitting of the pocket and blue jean collar.

Pre-war Jumper, Drill, worn as No. 6 and No. 7 Dress by Class II Ratings. The cuffs and lower edge are bound in blue jean, and the bottom edge of the jumper is fitted with twin eyelet holes, with a reinforcing patch at each seam. Issue of the drill uniform was suspended for the duration of the war in October 1939.

The single pocket fitted to the left inside breast of the Jumper, Drill. Also shown are the three rows of white edge tape hand stitched to the integral blue jean collar.

The basic Submarine Detector non-substantive badge pattern X86B. The badge consists of a harpoon and rope coil crossed by a lightning strike. In February 1940 a confidential order banned the wearing of SD badges to help prevent the enemy identifying Asdic specialists should they be captured. In May 1940 SD Ratings were authorized the wearing of the torpedo branch non-substantive badge.

Substantive badge pattern 3B for a Leading Rating worn over a single pattern 42B good conduct badge indicating three years' good conduct. The badges are blue on white, for wear on No. 6 Dress. Good conduct badges were introduced in 1849.

The crew of anti-submarine vessel No. 1 gives three cheers following commissioning at Chicago, in July 1943. The majority of the Ratings wear the Jumper, Duck, as Tropical Dress, but a few wear the pre-war Jumper, Drill, with blue cuff and hem binding and integral collar.

Interior pocket arrangement of the pre-war Jumper, Drill, with U front, showing detail of the pocket fitted to the right side and the becket for fitting the tape used to secure the silk.

A stoker wearing the tropical white No. 7 Dress with the pattern 26B blue on white Stoker 1st Class badge on his right sleeve.

The Jumper, Duck, worn with the separate blue jean collar by a Rating from the destroyer HMS *Kandahar*. Although it was required with this Dress, the neck lanyard has been omitted and the free end of the silk has been tucked into the V of the jumper.

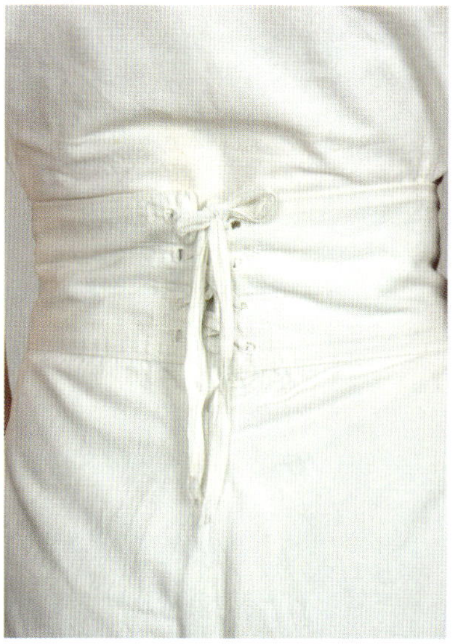

The rear waistband of pre-war Class II drill trousers showing the four-hole adjustment lace. The lace adjustment was used on all early 1930s Class II drill, duck and serge trousers, but was not seen on wartime garments. Clothing Regulations still listed lacing into the 1950s.

Paper label from a 'Jumper, Duck'. C. H. Bernard were a major contractor of RN uniforms. Size 17 was one of twenty-five individual sizes available for duck jumpers.

Pre-war Jumper, Drill with U front, made without the integral blue jean collar, necessitating the use of a separate collar. The garment is poorly constructed, possibly indicating a shipboard 'Make and Mend' garment. This type of garment was locally authorized on the New Zealand station, but it was also popular in the Mediterranean.

The Jumper, Duck, Working, worn with the blue jean collar in lieu of the No. 6 Dress Jumper, Drill, Uniform during World War II. The jumper's integral collar was occasionally cut away when the duck jumper was worn with a blue jean collar, making it cooler to wear.

Singlet, White Cotton, with Spinal Piece

This designation includes the Singlet, Cotton, Tropical, and the Singlet, Tropical. The Singlet, White Cotton, with Spinal Piece was made of white cotton with a simple oval neck opening and elbow length sleeves. A spinal piece was fitted to the back to assist in the prevention of heatstroke by insulating the spine. It was issued to all Class I, II and III Ratings until 1933 when a new open-neck shirt with collar – the Shirt, Tropical – was introduced for Class I Ratings.

Medical opinion on the benefits provided by spinal pads as protection from heat was divided during the inter-war period. In June 1937 a report by the Admiralty's Medical Director-General stated 'the spine pad is completely useless as an extra protection against the sun's rays, and only adds to the cost and weight of the article; it should be abolished.' In July a reply by the Director of Victualling noted the MDG's remarks but stated that 'there appears to be a considerable predilection for the spine pad among the

A photo taken in the Middle East during the mid-1930s showing the Class II Sun Helmet worn with cap ribbon, and the pre-1938 Tropical Singlet worn by the two men in the rear.

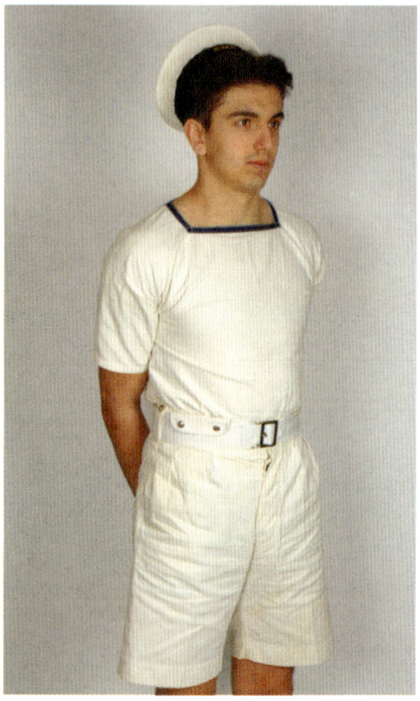

Seaman No. 10 Dress. This Rating wears tropical clothing consisting of the Tropical Singlet, white shorts, white waistbelt and white duck cap. Blue stockings were prescribed wear for Class II No. 10 Dress, but white were frequently worn.

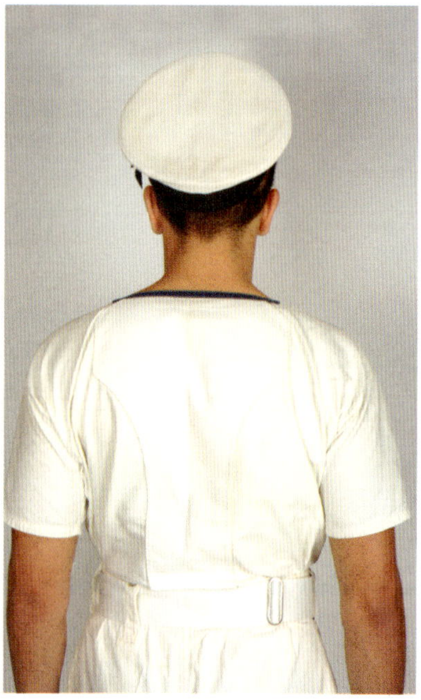

The back of the Tropical Singlet introduced in 1938, showing the spinal piece. The cotton flannel was frequently worn instead of the Tropical Singlet. The early pattern shorts had three belt loops, one at each side and one on the rear left. No loop was fitted at the rear right side as it would interfere with the sliding adjustment bar of the waistbelt, as shown here.

HMS *Repulse* in Haifa, Palestine, 1938. The sentry wears the Ratings' Sun Helmet with the HMS REPULSE tally, cotton flannel and drill trousers, an unofficial locally approved combination known as 'half whites'.

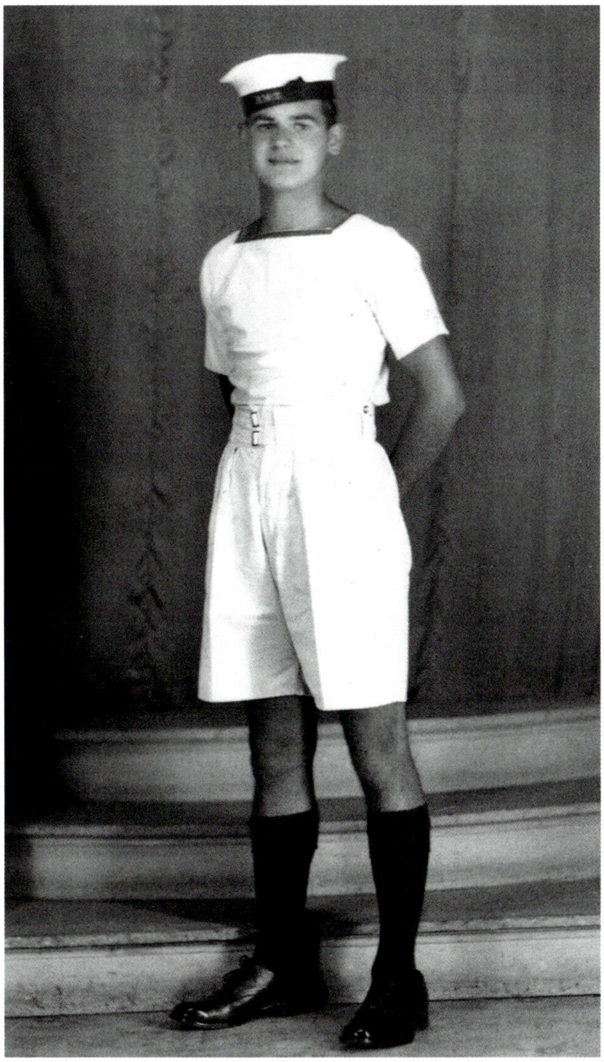

Studio picture of a Rating wearing No. 10 Dress, consisting of the White Duck Cap, Tropical Singlet, white shorts and blue stockings worn with black shoes.

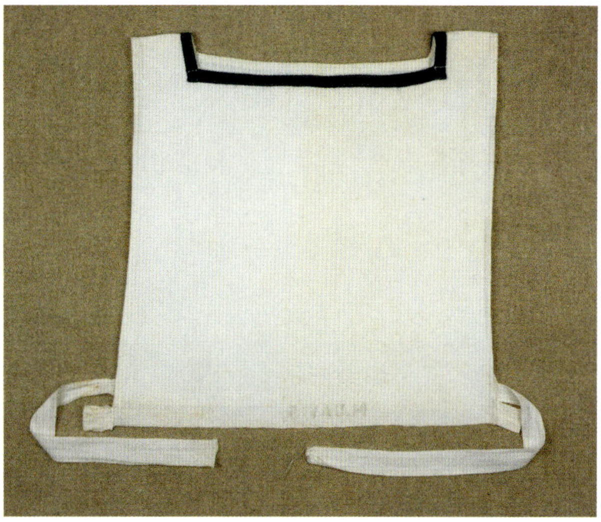

The standard cotton flannel was supplemented by the non-regulation but popular dickey style. It was worn beneath the serge, drill or duck jumper. In use since the early part of the century, the dickey was a product of the on-board 'make-and-mend' tailors and naval outfitters.

personnel of the fleet'. So despite the Medical Director-General's views, the spinal pad remained.

In 1938 the Singlet, White, Cotton, with Spinal Piece was replaced with a new design based on the cotton flannel. The new garment was essentially the cotton flannel but with a spinal piece fitted to the back. It was not uncommon for the cotton flannel to be worn instead of the tropical singlet as it proved to be both lighter and cooler. Despite the change of pattern the designation remained the same.

Dickey, Cotton, and Dickey, Flannel

The standard wool flannel, and the cotton flannel that replaced it, were both somewhat warm when worn with drill or duck jumpers, and were thus supplemented by the non-regulation but popular dickey-style singlet. In use since the early part of the century, the dickey was a product of the on-board 'make-and-mend' tailors and naval outfitters.

Numerous requests were made under the 1936 uniform review for the dickey to be officially manufactured and sold in Slops, and for the ship's tailor and trade versions to be recognized as authorized dress. The request was declined, but it did not prevent the dickey from being made on board ships or being sold by tailors.

Belt, Waist, Blue (Waistbelt, Blue)

Introduced in 1924, the 2¼in wide blue waistbelt was adjustable to size using a sliding toothless buckle. It had a small pocket on the right side with a flap top and two press-snap fastenings, and an integral knife suspension clip on the right side. The front was secured by a metal buckle and adjustable leather tongue. The belts were provided in two adjustable lengths, 40in and 46in. The issue of the belt was extended to RNR and RNVR seamen Ratings in August 1930.

In January 1935 the belt was modified: the knife clip was removed and the leather fittings were replaced by blue webbing. The loss of the knife clip resulted in an order that the clasp knife was in future to be worn on a lanyard around the waist by all seamen Ratings. However, many Ratings chose to fit a knife clip to the adjustment slider of the new pattern belt.

Belt, Waist, White (Waistbelt, White)

The blue belt was found to stain white uniform. To alleviate this problem AFO 960/39 introduced a white belt for optional use with white clothing. The buckle tongue had a single fastening position but was otherwise the same design as the blue belt. The belt was to be worn by men dressed as seamen with No. 5 or No. 7 Dress without jumpers. Men not dressed as seamen could wear the belt after working hours with the white shirt without a tunic when not on duty.

Austerity pattern belt

In 1945 an austerity pattern belt was introduced. It had a simplified buckle and a money pocket on the left side. This economy variant was not popular, but remained on sale alongside the old pattern for many years after the war.

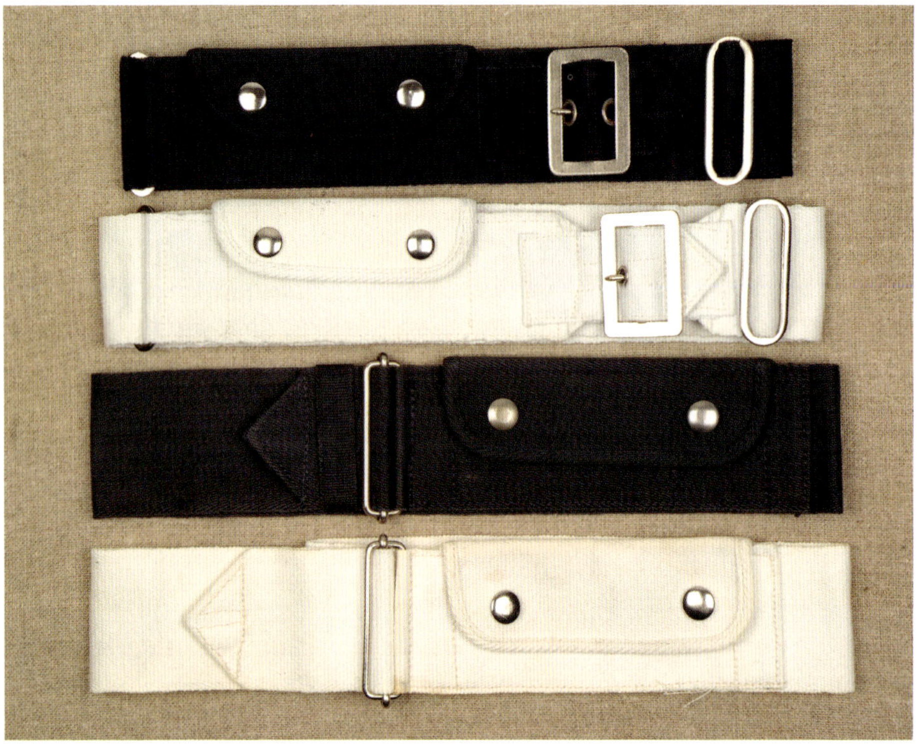

Class II waistbelts. The blue belt with money pocket was introduced in 1924 and modified in 1935. The blue was found to stain the white uniform, so a white belt was produced in 1939 for wear with the tropical clothing. *Below* are the war austerity patterns of belt introduced in 1945.

Overcoat

The Class II Ratings' Wool Overcoat was provided with an eight-button front – two rows of four with four to fasten. The collar was normally worn open but could be closed if required. Two small buttons were fitted at each cuff. Buttons were gilt for CPOs and POs, horn for others. The overcoat was modified in 1937, an adjustable half belt and waist hooks (to support the web belt when worn) were added, and the cut was improved. The adjustable half belt and cuff buttons were removed during World War II as an economy measure.

FAR LEFT: A seaman wearing the Class II Wool Overcoat with woollen mittens provided by the Royal Navy comforts fund. He wears a printed HM MINESWEEPER ribbon on his cap.

LEFT: The Class II Ratings' Wool Overcoat. This war-economy pattern has a fixed half-belt that requires stitching to the correct fit, and the cuffs are void of buttons.

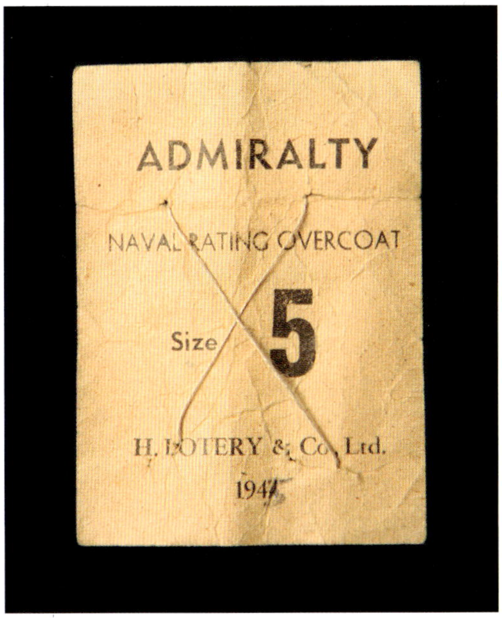

ABOVE: A 1944-dated label on a Rating's wool overcoat. The label is tacked on with an X stitch rather than being fully stitched as was practice on all prewar and early war garments. The 1944 date has been pencilled-over to read 1945.

RIGHT: Crewmen from HMS *Ajax* parade in London following the Battle of the River Plate. The men wear the Wool Overcoat with slung respirators.

Cap, Cloth, Blue; Cap, Duck, White

An Admiralty report produced in 1930 had shown that the regulation cap was highly unpopular. In the year ending March 1930 only 4,651 Class II blue cloth caps had been purchased from Slops, while by comparison 9,395 Class I and III peaked caps had been purchased. It was suggested that as many as 90 per cent of Class II Ratings were wearing caps purchased ashore from trade. But the Admiralty were already working on the problem, and in 1931 a new style of Class II white duck cap was adopted for all men dressed as seamen, having been approved in February 1930. The new cap had no separate stiffening grommet, but incorporated an improved lining, and was firmer than its predecessor, the integral fabric stiffener being carried to the edge of the crown. The following year AFO 382/32 introduced a blue cloth cap of the new pattern.

The introduction of new caps saw the issue of a duck cap cover discontinued, stocks of which were to be returned to victualling yards for disposal. In lieu of the duck cap cover Ratings who were required to wear a cap cover alone as a head covering were allowed to use a Marcelline cap cover as headwear. Such covers were typically worn by stokers in the old coal-powered steamships and by men employed on dirty work where the seaman's cap was not suitable.

The uniform review of 1936 had highlighted a number of complaints about the new caps and provided suggestions on how they could be improved. These included increasing the height of the cap and improving the stiffening to prevent it losing shape when wet. From the comments received from the men it was implied that the tradesman's cap was of superior quality to the issue version. To examine this, the review board purchased one duck and one blue cloth cap from a naval outfitter and a sample of service issue caps. The tradesman's caps cost 2s 5d each, whereas the issue caps were 2s 2d for the blue and 1s 7d for the duck cap.

All the caps were deconstructed and examined for quality of construction, fabric and finish. While the issue caps had a thin layer of waterproof cloth between the outer fabric and liner (to prevent sweat staining the outer fabric), the trade caps were found to be fitted with a sheet of brown paper between the outer fabric and liner. All the cap stiffeners were soaked in water for two hours. At the end of this time it was found that the service issue cap stiffeners retained their shape far better than those fitted in the trade caps. The trade caps were found to be slightly higher than the issue caps, and it was noted that they had liners of 'a very bright and striking colour'. This was usually in the form of a red and a green section of fabric, implying port and starboard sides, and gold-stamped manufacturer's logo typically accompanied by an impression of a Dreadnought or a submarine. Overall it was concluded that the service issue caps were 'greatly superior in every way' to those supplied by the trade.

As a result of the review a decision was made to slightly increase the height of the issue cap, and a recommendation made that trials should be carried out with a view to replacing the circular cap with one that was oval in shape, better conforming to the shape of the head. It was felt that the round shape of the cap contributed heavily to the tendency of Ratings to wear the cap 'flat-a-back' (on the back of the head) and that a change of shape to an oval form would reduce this inclination.

As with many of the review recommendations, the onset of war interrupted any further progress and it was not until just after the war that a Class II cap with an oval shape was finally made available. In September 1945, AFO 5563 introduced a new oval-shaped seaman's cap that was considered more comfortable to wear than the old round pattern. The crown of the new cap was $1\frac{1}{4}$in longer in the longitudinal axis than it was in the lateral axis. Accumulation of stocks of the new cap was very slow, and blue caps of the old round form were still listed in stores books alongside the new oval caps as late as 1952.

It was a regulation requirement that the white cap was worn with white clothing, and it was also to be worn as a summer cap between 1 May and 30 September in home ports. The blue cap was to be worn at other times. Compulsory holdings for men on Home Service were two blue caps and one white cap between 1 October and 30 April, and two white caps during the summer period 1 May to 30 September. For men on foreign service, the compulsory holding was two white caps with one blue cap being optional.

The uniform review of 1936 received suggestions that the blue cap should be abolished and that the white cap be worn on all occasions year round, but the board were strongly against this and maintained a requirement for both blue and white caps. Interestingly the outbreak of war saw a change quite the contrary to the men's wishes: AFO 1307 of April 1940 instructed Officers and Ratings that white caps or white cap covers were no longer to be worn by Officers and men on shore in the United Kingdom or on board ships in home waters for the duration of the war. In April 1946 white caps or white cap covers were once again authorized for wear by Officers and men on shore in the UK or on ships in home waters between 1 May to 30 September as per Uniform Regulations. The blue cloth caps were finally abolished in 1956, some nineteen years after it was first requested that white caps be worn all year.

ABOVE: *At left* is the post-1931 summer cap with duck top and high crown, and *at right* the post-1932 winter cap with high crown and blue cloth top.

LEFT: An MTB crewman wearing the White Duck Cap with H.M.M.T.B. (His Majesty's Motor Torpedo Boat) cap ribbon.

Buglers from HMS *St Vincent* wearing No. 5 Dress and blue cloth caps during the autumn of 1933. The boys wear a mix of pre- and post-1932 cap patterns. The new caps show a much better shape, are stiffer, and have a higher crown compared to the rather shapeless older pattern.

ABOVE: A view of the inside of post-1932 Class II caps. *At left* is the regulation type in white duck; *at right* is a commercial 'tradesman's' blue cloth cap for winter wear. The tradesman's cap has a red and green lining and a gold crown stamp showing a Dreadnought; other designs included a Submarine. The round form of World War II pattern caps is evident in this view.

LEFT: A Class II Rating photographed in 1940 wearing the HM DESTROYER cap ribbon on the Cap, Duck, White.

ABOVE: The oval shape cap introduced in September 1945 under A.F.O. 5563/45. The white duck cap was also produced in the new oval shape. This pattern provided a better fit and was more comfortable than the round wartime pattern, and also helped to prevent the cap being worn 'flat-a-back'.

RIGHT: A trainee communications Rating wearing the Cap, Cloth, Blue with H.M.S. IMPREGNABLE cap ribbon.

Box, Cap

In October 1920 the cap box was reintroduced to the kit of men dressed as seamen. It had been removed in March 1917 because of a shortage of metal. The box was made of thin sheet steel, large enough to hold two caps, with a lid and closing latch that could take a small padlock. The cap box was typically used to hold one blue cloth cap, one white cap, 1 blue collar, housewife, toilet gear, writing utensils and small personal effects. As an economy measure during World War II the cap box was once again removed from compulsory kit lists, and it remained as a temporary loan item for boys under training well into the 1950s. However, the box was a popular item with men who already held one, or could beg or barter for one, as it provided a secure storage for personal items.

Cap boxes. *At left* is the standard World War II pattern 134A cap box, and *at right* is the old-style box that it replaced. The new box was ¾in higher to accommodate the post-1932 pattern caps. Both have a brass plate fitted to the top bearing the owner's name and initials.

In 1932 the height of the cap box was increased by ¾in. This increase was required as it had been found that caps purchased privately from naval outfitters, officially termed 'tradesmen's caps', were deeper in the crown than service issue caps; also the caps from service sources were being modified to have a slightly increased height, and when two such caps were stored in the cap box the lid would not close easily.

Ribbons, Cap, Lettered.

The gold wire on silk-type cap ribbon (cap tally) was introduced in 1868. Prior to this date the ship's name had been hand painted on to ribbons or caps in a variety of ornate and non-uniform styles. In February 1918 AWO 662 introduced yellow silk lettering on cap tallies as a wartime economy to conserve stocks of gold wire used in the lettering of tallies. The use of yellow silk lettering was quite short-lived as AMO 2665 of 6 August 1919 reintroduced gold wire lettering.

AFO 226/24 of January 1924 stated that in future all cap ribbons worn by the crews of submarines were to bear the individual designation of the respective submarine instead of the H.M. SUBMARINES ribbon previously used. A quantity of H.M. SUBMARINES cap ribbons should be retained for use by spare (unallocated) crews. A variant of the submariner's tally bears the lettering H.M. SUBMARINE.

Kings' Regulations and Admiralty Instructions stated that cap ribbons were issued at a rate of two per man in home waters and three for those on foreign stations upon joining a ship or establishment. The third tally was for use with the Sun Helmet. Ratings taking passage on HM ships were issued a single ribbon for that vessel, although this practice was soon stopped as an economy measure during the war. Naval Uniform Regulations stated that the cap ribbon lettering was to be ½in in height, with a ribbon of black silk 1³⁄₁₆in wide and 45in long. This length allowed the ribbon to be used on both the Class II cap and the Sun Helmet. As a wartime economy measure the cap ribbon was shortened to 37in in October 1940, after which time it was ordered that the ribbon's free ends were to be stitched together without a bow when worn on the Sun Helmet. The reduction in length was to save in the region of 100,000 yards of ribbon a year. The use of cap ribbons on the Sun Helmet was discontinued altogether in 1943.

Plain H.M.S. tallies without any ship's name were in use during the Great War. They were re-introduced to the RN in March 1939 for use by boarding boats' crews, the requirement being brought about following Royal Navy operations during the Spanish Civil War. H.M.S. tallies were to be demanded by ships that might be required to carry out boarding operations and examination duties. Cruisers and above were to hold fifty tallies, while smaller vessels were to stock twenty tallies for boarding parties. The ribbons were to have silk or gold wire lettering and were to be issued for boarding operations only, being withdrawn after use.

The use of plain H.M.S. ribbons was extended in January 1940 under C.AFO 64 (Confidential Admiralty Fleet Order 64). All seamen proceeding on long leave from battleships, battle cruisers, cruisers and destroyers were to be loaned plain H.M.S. tallies for the duration of their leave, returning them on rejoining their ship. Men proceeding on draft to join new ships were also required to wear the H.M.S. tally when in transit. The tally was to be exchanged for a new named ribbon when joining their new vessel. H.M.S. tallies were to be issued to men serving on fleet auxiliaries such as trawlers and other small craft, as well as on shore establishments where manning levels were not sufficient to justify the expense of producing named tallies. New entry 'hostilities only' Ratings were to wear H.M.S. tallies whilst under training. The use of plain H.M.S. tallies in other circumstances could be ordered by the Admiralty or Commander-in-Chief, but the substitution of named tallies was not intended.

In the spring of 1940 a new cap ribbon was authorized for use by members of the Patrol Service. This service manned a variety of small craft ranging from trawlers, tugs and drifters to yachts. The new ribbon bore the lettering 'H.M. PATROL VESSELS'.

By the end of 1940 a severe shortage of materials had led to the individual issue of two tallies (three for foreign service) temporarily being restricted to a single cap ribbon at home and abroad for all Ratings, and the suspension of issue of any ribbons bearing lettering other than H.M.S. once the stocks of other ribbons had been used up; the order included H.M. SUBMARINES and H.M. PATROL VESSELS ribbons.

It was not uncommon to see a mix of different named ribbons being worn on board ships, as men joining a new vessel frequently retained their old ribbon, often reluctant to change if the ribbon bore the name of a famous or capital ship. By the close of 1942 the availability of tally stocks allowed the return to the previous scale of issue of two tallies to all new entries, with a third tally being issued on being supplied with a Sun Helmet (the use of tallies on Sun Helmets was discontinued for the duration of the war in July 1943 under AFO 3214, the issue of a third tally for use on the Sun Helmet being stopped).

The use of ship's names on cap ribbons gave rise to concerns over security, as enemy agents operating in ports would know which vessels were alongside by the cap ribbons being worn. In September 1942 AFO 4678 announced that despite stocks still being available, individual named ribbons were no longer to be issued during the war and that in future only ribbons lettered H.M.S. would be provided. From that time only H.M.S. ribbons could be worn ashore. However, the use of named ribbons was allowed to continue until the 31 March 1943 for men at sea, after which only plain H.M.S. cap ribbons were to be worn in the Fleet.

The rules governing the use of named ribbons had applied to HM submarines, and by March 1943 named ribbons were no longer in use. However, while named ribbons had been abolished, men in possession of H.M. SUBMARINES cap ribbons could continue to wear them until they were worn out. While no new H.M. SUBMARINES ribbons were to be manufactured, those available in submarine depot ships could still be issued until stocks were exhausted. In lieu of the H.M. SUBMARINES cap ribbons Ratings serving on submarines were authorized the use of an economy pattern ribbon bearing only the lettering H.M. S/M under AFO 3481/43.

As well as the official Admiralty issue cap ribbons a number of commercial patterns were available for purchase from naval outfitters. Typical examples included the lettering H.M. DESTROYER, H.M. CRUISER, H.M. MINESWEEPER and H.M. CONVOY, but many other variations were available. Pre-formed 'tiddley' bows ready to be stitched on to a cap ribbon could also be purchased from retailers. Occasionally a Rating chose to have a removable bow secured by a snap fastener, and others embroidered over the lettering of the ribbon. Early commercial manufactured ribbons were standard pattern gold wire on silk, but as austerity measures and rationing came into force manufacturers turned to rayon lettering and then gold print on rayon ribbon. Official Slops-issue ribbons also suffered from the shortage of silk, being made wholly from rayon from 1942.

The *Seaman's Pocket Book* (BR 827) stated that 'The cap ribbon must be worn so that the "H.M.S." is directly over the nose. It must be tied in a reef bow over the left ear.' AFO 154/44 reminded Ratings that cutting cap ribbons, sewing the ends together and then adding a separate bow was against regulations. The ends of the bow were to be of equal length, not exceeding 2in. In practice 'tiddley' bows often exceed those dimensions. The change from 45in to 37in ribbons meant that to wear the bow tight alongside the H.M.S., as was common practice, meant that the ribbon had to be cut and stitched, as the length of the free ends of the 37in ribbon meant it was not possible to tie the ribbon with the bow in such a position. A second ribbon was needed to make up the separate bow, or one could be bought from retailers. Wartime images show that BR 827, AFOs and uniform regulations were universally ignored, with the bow almost always being tight up to the H.M.S. and worn over the left eye.

Ratings who were accepted as Officer candidates and who were undergoing training for their commissions were authorized the wear of a plain white cap band with no lettering. AFO 2888 of January 1940 authorized the issue of two 'Bands, Cap, White' to each trainee Officer. The bands were worn by all Ratings, including those in Class I and III uniforms, who wore the ribbon over their badge, or on occasion, fitted their usual cap badge to the ribbon.

In February 1946 AFO 1441 reintroduced tallies lettered with individual ships' names, except for submarine and minesweeper crews who were to be provided with tallies lettered H.M. SUBMARINES and H.M. MINESWEEPERS respectively. Lettering was to be yellow rayon on a black rayon ribbon. It was made very clear that production of the new named tallies would be a slow process due to extreme shortages of materials and manpower. As late as 1948 the supply situation was still severe and H.M.S. tallies were in widespread use as the production of named tallies had failed to meet demand.

In 1951 AFO 2309 reintroduced gold wire lettering. From July, Ratings were to receive one gold wire and one rayon-lettered cap tally on entry or change of ship. At this time plain H.M.S. ribbons were still in use, including issue at some shore training establishments, men taking passage in HM ships, and men on short training courses.

Gold wire ribbons bearing generic designations. *Top to bottom*, FLEET AIR ARM, H.M. PATROL VESSELS, H.M. MINESWEEPER and H.M. DESTROYER.

Crewmen from an unidentified submarine wearing typical hot weather clothing, except for the officer at left, all without shirts. While the majority of the Class II Ratings wear HM S/M cap ribbons, a couple wear HMS and a single Rating wears the H.M. SUBMARINES ribbon.

Cap ribbons bearing the names of RN warships showing typical early war lettering: H.M.S. CURACOA, light cruiser sunk October 1942, H.M.S. RAMILLIES, battleship scrapped in 1949, H.M.S. SHROPSHIRE, cruiser transferred to RAN in 1943, H.M.S. ARK ROYAL, sunk 1941. The CURACOA ribbon has a particularly fancy bow.

Wartime cap ribbons bearing only the lettering 'H.M.S'. *Top to bottom*, a printed ribbon purchased from commercial sources, an issue ribbon in gold wire, an issue ribbon in gold rayon, and an economy ribbon woven on a stitched-edge tape.

LEFT: The 'ROYAL NAVAL RESERVE' tally worn on the blue cloth winter cap by a Rating reservist photographed during 1936.

RIGHT: The Cap, Cloth, Blue worn by a Class II Rating. The cap bears the standard 'H.M.S.' tally with a 'tiddley' bow stitched to the ribbon close up to the lettering.

POs and other Ratings wearing the plain white cap band, indicating that they were undergoing training to receive commissions as Officers.

TOP LEFT: An H.M. MINESWEEPER cap ribbon worn on the blue cap by a General Service Rating.

TOP RIGHT: A Class II Rating wearing a printed FLEET AIR ARM ribbon on his blue cloth cap.

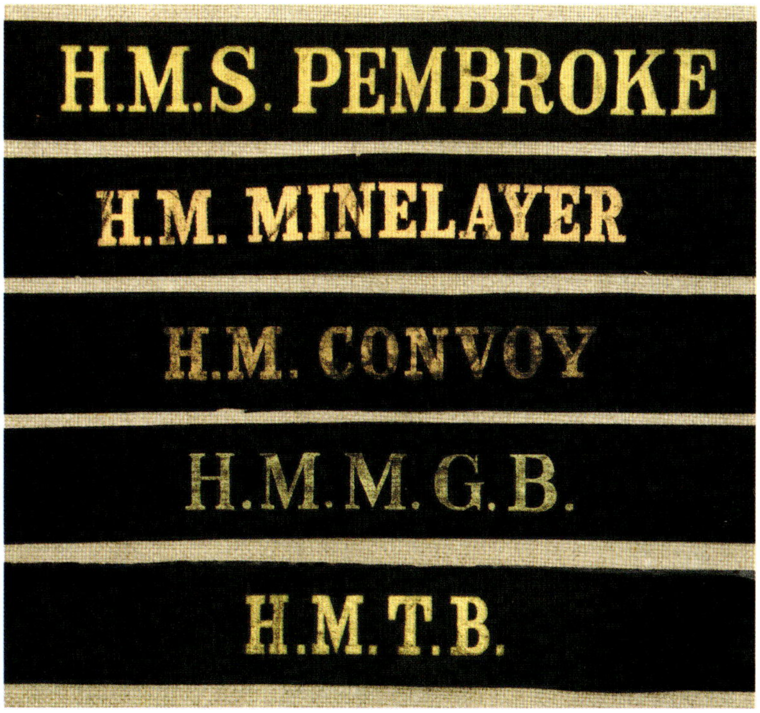

Typical examples of printed ribbons privately purchased from commercial sources. H.M.S. PEMBROKE, a shore establishment at Chatham, H.M. MINELAYER, H.M. CONVOY, H.M.M.G.B. (His Majesty's Motor Gun Boat) and H.M.T.B. (His Majesty's Torpedo Boat).

3 Battledress, Khaki Drill and Jungle Green Uniform

Combined Operations and RN Commandos

RN involvement in Combined Operations (CO) initially involved only a requirement for landing craft crews, but it was soon realized that naval beach parties were essential for amphibious operations. The beach parties provided specialist support for landing units, communications, beach and obstacle clearance. The first use of naval beach parties was during the landings on Diego Suarez in May 1942. The success of the RN's involvement led to the rapid expansion of the CO units and RN Commandos, often called 'Beach Commandos'. Dieppe saw the first fully integrated use of RN Commandos, when C and D RN Commandos provided Beach Masters for each of the beaches to be assaulted. Dieppe was a disaster, and the RN Commandos suffered heavy losses, but much knowledge was gained. This knowledge proved invaluable for the Operation *Torch* landings in North Africa in November 1942, and again during Operation *Husky* landings on Sicily in July 1943.

The landings at Salerno and Anzio showed without a doubt the value of the RN Commandos, and by the time of the Normandy landings in June 1944 the units were well prepared for a major amphibious operation and were an essential component of Combined Operations. The Allied move into mainland Europe did not mean an end to RN Commando operations, and naval Commandos were to see action at Walcheren and the Rhine crossing; most RN Commandos were then sent to the Far East in preparation for the invasion of mainland Japan. The RN Commandos had already been involved in operations in the Far East, including the Arakan offensive in 1944 and Malaya in early 1945. However, the dropping of the atomic bombs negated the requirement for a mass and costly assault on the Japanese mainland, in which the RN Commandos were to have had a pivotal role.

As well as RN Commandos, a number of other units served under Combined Operations, receiving military training pertinent to their role. These units included major landing craft, minor landing craft, Naval Beach Signal Sections, Naval Sections Bombardment Troops, Landing Craft Obstruction Clearance Units, Landing Craft Recovery Units, Landing Ship Tank (LST) crews and Pontoon Barge Units.

It had been appreciated from the outset that men serving with Combined Operations needed specialist military clothing rather than standard naval dress. This requirement was met by the issue of army-pattern clothing and equipment. A basic set of army kit was issued gratuitously to all CO Ratings: the Blouse, serge; Trousers, serge; Shirts, Angola, drab; Blouse, denim; Trousers, denim; and Socks, thick, grey, were the basic components of the army battledress uniform. It was considered that the battledress issue corresponded with the following naval items (or the corresponding uniform item appropriate to Rating): Jumper, Serge; Trousers, Serge; Shirt, white; Suits, combination, jacket; Suits, combination, trousers and Socks, thick, blue, and CO clothing was priced accordingly in Slops. As gratuitous issue kit when CO items required replacement, they had to be purchased using the Ratings KUA Officers had to purchase their own CO clothing and could do so through the naval stores system.

The units that were authorized special Combined Operations clothing and equipment issues were the crews of major landing craft, minor landing craft, Naval Beach Commandos, Naval Beach Signal Sections, Naval Sections Bombardment Troops, Landing Craft Obstruction Clearance Units, Landing Craft Recovery Units, Landing Ship Tank crews and Pontoon Barge Units. The Combined Operations scale of equipment was referred to as 'CO Kit'.

By early 1945 the changing nature of operations in which naval ranks were involved had brought about a major review of equipment issued to Combined Operations personnel. The anticipated defeat of Germany and an emphasis on moving all available resources to defeating the Japanese had brought about a greater requirement for specialist jungle clothing for Combined Operations Officers and Ratings operating in the East Indies. Special scales of issue of Jungle Clothing were promulgated in AFO 1955/45. The AFO made mention of new patterns of olive green bush jackets, trousers and hats as being in production, and that they would be issued as stocks allowed (this AFO referred to the 1944 pattern Jungle Clothing and equipment). Jungle green clothing was already being produced locally in the East Indies, often merely KD clothing that had been dyed green, and was being issued to Officers and Ratings in theatre as required. The use of jungle clothing was solely as a special duties working or action dress, and it was not to replace standard naval uniform. However, Officers and Ratings serving with CO were not required to hold white tropical Dress, No. 13 or No. 10, for which they could wear Khaki Drill (KD) or Jungle Green (JG) in lieu.

The scales of equipment issued to the different components of the CO varied depending on the individual's employment. As per usual naval practice the gratuitous issue items were to be replaced at the owner's expense, but loan clothing items were exchanged one-for-one, as required, at no cost to the individual. Web equipment and weapons were unit stores, as were mosquito veils, nets and gauntlets, and were not treated as a loan issue. The items of equipment authorized for issue to RN Beach Commandos are representative of the general CO kit scales and included the following:

Webbing: Belt, waist; bottle, water; carrier, bottle, water; braces, webbing; frog; haversack; pack; straps, pack; pouches (ammunition); attachment, brace; sling, rifle; straps, shoulder; implement, entrenching, head; implement, entrenching, helve; carrier, entrenching tool. Pistol equipment was issued to Officers and Ratings who were required to carry a pistol.

Standard CO Naval pattern kit: Rope-soled shoes, Royal Navy shoulder titles, CO badges, knitted blue woollen cap, light transport bag, steel helmet anti-gas curtain and anti-gas oilskin gloves.

Standard CO Army pattern kit: Blanket, army pattern boots, web anklets, leather jerkin, serge blouse (BD), serge trousers (BD), trouser braces, khaki shirts, grey socks, denim blouse, denim trousers, groundsheet, knife, fork and spoon, tin plate, enamel mug, holdall, field dressing, helmet camouflage net, helmet comforter, personal camouflage net and anti-gas cape.

Tropical CO Army pattern kit: KD Bush jacket, KD trousers, KD shorts and khaki hose tops.

East Indies CO Army pattern items: Jungle gaiters, jungle boots, heavy, jungle boots, light, jungle bush jacket, jungle trousers, jungle underclothing, jungle socks, jungle towel, jungle hat, groundsheet cape, face camouflage net, petrol lighter, jungle shorts, jungle hose tops. This list refers to the 1944 pattern Jungle Equipment, little of which reached the Far East prior to the cessation of hostilities. The majority of JG clothing worn was of Indian manufacture, often standard KD uniform dyed JG.

During assault training in Scotland a Royal Navy Leading Seaman wears army Battledress with blue cloth cap and 1937 pattern web equipment.

The left side badge of the paired Pattern 291A Combined Operations badges authorized under AFO 4366 of September 1942. The badges were worn by all RN personnel attached to CO and wearing Battledress.

Sub Lieutenant RNVR shoulder straps modified for wear with Battledress (BD). The straps have had loops of cloth fitted to the reverse so they may be easily fitted to the BD.

LEFT: A Sub Lieutenant of the RNVR attached to Combined Operations. He wears army Battledress fitted with the Combined Operations badge and ROYAL NAVY shoulder titles.

RIGHT: A fine study of an RNVR Lieutenant wearing Battledress. He has ROYAL NAVY titles at the shoulder and Khaki Drill slip-on rank slides.

Arctic CO kit: Waterproof smock, light oilskin trousers, submarine sweater (frock), knitted slippers, comforter, spun neck cloth, balaclava helmet, fleece-lined jerkin (replacing the sheepskin coat), rubber Seaboots, Seaboot stockings, sleeping bag, army type, fleece-lined gloves and light transport bag. Arctic clothing and equipment was only to be issued on receipt of special orders.

Note: These CO clothing and equipment lists were amended on a number of occasions between first drafts of 1942 and those in existence at the end of the war. The above are representative of the items listed for issue in late 1944, although not all items were actually available for issue at that time.

Combined Operations badges: For CO badges, see Substantive, Non-Substantive and Miscellaneous Badges under the Miscellaneous Clothing and Equipment section.

Battledress

Army BD had been approved for use by naval ranks serving in beach communication parties by C.AFO 2079/42. Apart from the beach comms' and a few select users, the use of army pattern BD by Naval Officers as a working dress was expressly forbidden at that time. However, the expansion of Combined Operations saw a massive increase in the number of naval ranks that were authorized the use of Army pattern BD.

Army pattern clothing, including Battledress, was supplied to the International Red Cross in Geneva for issue to prison camps within Europe in which British Prisoners of War were held. The list of army pattern clothing supplied to the Red Cross for distribution to POWs included Battledress blouse, Battledress trousers, Great Coat, Pullover, FS Cap, ankle boots, Angola shirts, underwear, socks, gloves.

The cost of any Army clothing issued to Royal Navy Officer POWs by the Red Cross was debited to their pay. However, if any items were returned to Government stores in a serviceable condition after repatriation, a credit would be given. A credit was also allowed for items of clothing that of necessity had to be abandoned in prison camps. A special form 'Naval Form P.O.W. Clo.' was given to each

A Royal Navy Beach Signals Rating in full army-pattern Battledress uniform. Only his insignia identifies him as Royal Navy.

Survivors of HMS *Gloucester* in an Italian POW camp. They wear Battledress with rate badges worn on the left sleeve. Field Service caps are evident, with at least two bearing the PO's cap badge. A CPO at far left has fitted his badge to the left breast pocket of the BD blouse.

RIGHT: Seamen Ratings on board an unknown warship. The sailor *at right* wears a post-1942 austerity pattern Battledress, with Royal Navy titles, Gunlayer 3rd Class and good conduct badge applied, army shirt and Class II serge trousers, a specifically unauthorized mixed rig.

The CO badge and Beach Signals flash worn by Officers and Ratings attached to the RN Beach Signals Section.

repatriated POW Officer, and all clothing received by them during their time as a POW had to be listed on the form so charges could be raised by their supply officer. As examples, Officers were charged 19s 9½d for a Battledress blouse, and were credited with 10s if the item was returned or declared abandoned. An FS Cap was charged at 2s 5d and credited at 1s 3d if returned or abandoned. The indignity of being charged for clothing issued while being held by the enemy as a POW can only be imagined.

Ratings RM SD

An interesting and secret circular 3392/40.Q issued by the Admiralty on 1 February 1940 stated that forty-four naval Ratings were to be issued with two sets of Royal Marine pattern Other Ranks Service Dress (without shoulder titles), two SD caps and one pair of puttees. It added that the SD was to be issued with RM buttons fitted, but that it was intended that these would be replaced by RN pattern in due course. No further information is available as to why the Ratings in question required RM SD uniform.

Khaki Drill Uniform

During the early stages of the war the tropical shirt (or tropical singlet) and white shorts had been the normal working dress of Ratings on all tropical stations, regardless of their duties. This had been found to be wholly impractical dress in many of the varied and far-flung locations that naval Ratings found themselves working in. The situation was remedied in July 1941 when the issue of KD clothing to all men serving at specified RN Air Stations and Air Sections abroad was authorized. Ratings drafted to Air Stations or Sections at Gibraltar, Jamaica, Trinidad, Bermuda, Freetown, Capetown, Dekheila, Fayid, Aden, Trincomalee, Morib and Sembawang were to be issued two pairs of khaki shorts and two khaki shirts in lieu of white tropical shirt or tropical singlet and shorts. Replacement on station was to be made from Army or RAF sources. Individuals could purchase khaki stockings and brown shoes for wear with the khaki shirt and shorts at their own expense.

Under 1944 regulations Royal Navy Ratings serving ashore in Australia were to be provided with two khaki bush shirts and two pairs of khaki shorts in lieu of the white tropical shirt or singlet and shorts. In malarial areas two pairs of KD trousers, two pairs of hose tops and a pair of web anklets were also issued. Naval Air Ratings were to be supplied all of the above items, and a khaki Sun Helmet. AFO 1954 of April 1945 notified that in future the gratuitous issue of tropical clothing (white or KD) for Ratings proceeding from the UK to stations where tropical clothing was worn were to be issued three shirts or tropical singlets (bush shirts or khaki jackets if KD was authorized) and three pairs of shorts per man. Fleet Air Arm Ratings were to receive two suits of white clothing and one of khaki before departing the UK. Upon arrival at their overseas station they were to be provided with an additional white suit and two KD suits, bringing their total issue to three of each.

By April of 1945 the large number of naval personnel who were authorized the wearing of KD uniform had resulted in a heavy demand on the available stocks. With the pending cessation of hostilities in Europe the demand would increase dramatically as more men and resources were channelled into the fight against the Japanese. Production levels fell far short of demand.

A Class II rating wearing Khaki Drill clothing consisting of KD Aertex shirt, with slip-on ROYAL NAVY titles, and drill trousers.

Detail of the ROYAL NAVY titles worn on the epaulettes. This is a private purchase, theatre-made item.

FAR LEFT: Period image of a Rating wearing Khaki Drill uniform with the white duck cap. The Rating wears the ribbon of the 1939–43 star, an unauthorized adornment on KD uniform.

LEFT: Ratings wearing Khaki Drill uniform in Italy, 1945. The leading Rate *at left* wears ROYAL NAVY shoulder strap badges.

A Royal Navy beach party operates a loudhailer during training near Kabrit, Egypt. The men wear standard army-pattern Khaki Drill uniform and brassards bearing the letters 'NBP'.

At a RNAS, somewhere in Africa, Ratings parade in a rather motley array of clothing and footwear, including truncated overalls, white and Khaki Drill shorts, tropical flannels and shirts, white duck caps, and white and khaki sun helmets.

Officer's KD Uniform

KD working uniform had been worn by Officers serving on the China station during the mid-1930s. The dress consisted of KD shorts, shirt, stockings and cap cover. Initially approved only for use on submarines, its use was eventually extended to other units on the China station.

Officers serving with CO or on air stations where KD was authorized wore KD clothing as working dress. In March 1945 Officers were advised that they could purchase Ratings KD clothing from service sources if they had been appointed to serve in stations where tropical clothing was worn. The items available were khaki bush shirts, khaki drill trousers, khaki shorts, web anklets and hose tops. It was not until May 1945 that Officers' uniform was enhanced by the official authorization of KD as a part of the tropical uniform. The uniform, based on that worn by the Army, consisted of khaki-coloured KD bush jacket, shirt, shorts or trousers and a KD cap cover. The KD uniform was to be worn with black or tan stockings and with either brown or black shoes. Shoulder straps were to be of the pattern used with No. 13 and No. 5A Dress, or alternatively they could be khaki with khaki braid rank lace.

In Australian waters No. 13 Dress was standard for Officers serving afloat. No. 10 Dress was to be worn for ceremonial and ashore at night unless white tropical No. 13 Dress shirt and shorts was authorized. KD tropical dress was authorized as working dress ashore. KD Service Dress of Army pattern could be worn in shore establishments when topical clothing was usually worn, or in lieu of white Undress uniform at night. Anti-malarial rig of long-sleeve bush jacket, trousers and hose tops was to be worn where malaria was endemic. These uniform items were to be provided before departing the UK. However, provision had been made for items to be manufactured and purchased locally in Australia if required. Officers already possessing tropical clothing items not conforming to regulations were permitted to wear them until they required replacement.

Prior to official Admiralty authorization for the wear of KD uniform by Officers, it had long been practice that RN Officers serving in the USA and Canada during the summer months wore khaki uniform. This was frequently of local purchase and of standard USN or Canadian pattern with RN buttons and rank added.

Web equipment, battledress, KD and JG uniform are illustrated and described in detail in this author's works *British Web Equipment of the Two World Wars*, *The World War II Tommy: British Army Uniforms European Theatre 1939–45* and *Khaki Drill and Jungle Green*.

A Lieutenant Commander RNVR wearing a Khaki Drill bush jacket and cap with KD cover. This jacket is fitted with gilt buttons, although KD was occasionally fitted with bronzed buttons.

A Sub Lieutenant RNVR stands alongside a 20mm AA mounting. He wears the Khaki Drill bush jacket with blue cap and white cover.

LEFT: Detail of the khaki lace on Khaki Drill RNVR Lieutenant's slip-on rank devices as worn with the khaki uniform. AFO 2471/41 stated that as a wartime measure Officers wearing khaki uniform could wear either blue shoulder straps with gold lace or khaki shoulder straps and lace

RIGHT: This RNVR Lieutenant wears the standard Khaki Tropical Dress uniform of bush jacket and trousers. The jacket is made from Aertex fabric, but cotton drill jackets were also used.

A Khaki Drill bush jacket with truncated sleeves. This jacket is fitted with eyelets for the attachment laces of the KD RNVR shoulder straps.

A lightweight cotton bush jacket in the style of the standard drill jacket. It has button cuffs and shoulder eyelets for attachment of rank straps, in this case a Captain (E).

RIGHT: Detail of the Captain (E) strap fitted to the cotton bush jacket. Originally Lieutenant straps, they have been upgraded following promotions by the addition of new lace, a typical wartime economy.

LEFT: An RNVR pilot DLCO (Deck Landing Control Officer), serving with 1820 Naval Air Squadron, US Naval Air Station Brunswick, 1944. The Lieutenant wears khaki service dress-style jacket with rank straps, khaki shirt and tie, and blue cap and khaki cover.

A US Navy Service Dress Khaki jacket used by an RNVR officer serving in the USA. A highly irregular dress, RN Officers' buttons have been fitted to the jacket along with RNVR Lieutenant laced US pattern rank straps.

Admiral the Lord Mountbatten photographed on the Arakan front during early 1944. He wears a Khaki Drill bush jacket, KD trousers, and cap with khaki cap cover. The cap's peak has the double oak leaf embroidery worn by Flag Officers and Commodores First Class.

ABOVE: A unique pair of rank straps for a Lieutenant RNVR. The straps have been made up using USN straps with integral US Navy eagle and anchor snaps. Such unorthodox uniform items are rarely encountered and are a significant departure from acceptable dress.

LEFT: Senior officers on the bridge of HMS *Empress*, May 1945. *Left to right* are Captain H. A. Traill, Vice Admiral H. T. C. Walker, Flag Officer commanding 3rd Battle Squadron, and Commander Briggs. Traill wears Khaki Drill shirt and shorts, Walker wears KD shorts and bush jacket (all regulation pattern), while Briggs wears No. 13 Tropical Dress, the shirt being of the Class I and III Ratings pattern.

4 Miscellaneous Clothing and Equipment

The fur hat had a quilted cotton lining with fur-covered side flaps that could be dropped down over the ears, and a fur-covered peak. The cap was issued for wear in extreme cold, such as on arctic convoys.

Flight-deck crew from HMS *Biter* dressed for extreme cold. Most of the men wear the fur hat. Stood at the *extreme left* is a handler wearing the inflatable lifebelt fitted with the pocket for the lifebelt lamp. *At right* the lifebelt is worn with the lamp taped directly to the neck strap. WWW.ROYALNAVYRESEARCHARCHIVE.ORG.UK

At left is the blue wool glove issued to Class II and Class III Ratings. *At right* the white wool glove issued to CPOs and POs. White wool gloves could be purchased and worn by Officers in cold weather.

RIGHT: Woollen gloves, mittens, comforters and balaclava helmets were issued gratuitously from service sources to personnel serving in northern waters. Comforts committees also provided 'gift woollens' to seamen.

LEFT: Admiralty-issue flannel 'Pyjama suit' as used by all Ratings. Officers were required to possess both heavyweight (flannel) and lightweight (cotton) pyjamas.

RIGHT: Royal Naval wartime-issue wool underpants and Long Johns. These are the standard 'WARNORM'-pattern woollens provided to all three services.

Two variants of the white scarf worn by Ratings in Class I and Class III uniforms with overcoats, oilskins or waterproof coats when on duty. All Ratings were allowed to wear the scarf while on leave, and it could also be worn with night clothing by CPOs and POs not on duty.

A young Class II seaman Rating photographed while on leave wearing the white scarf as a fashion accessory.

ABOVE: Fleet Air Arm fire fighters in RAF-pattern asbestos suits. The suits allowed the men to enter fire with minimal risk of burn injury from the flames. The suits were invaluable in rescuing aircrew from burning aircraft.

RIGHT: Detail of a group photograph showing fire crew sat on a tender at HMS *Tern*. The men wear Fearnought jackets, flameproof garments used by fire fighters and stokers.

Physical Training Instructors and sports teams wore a PT rig consisting of white shorts, white shirt with blue trim, and a blue elastic 'snake belt'. The PT rig is worn here by Ratings undergoing a boxing training course during 1935.

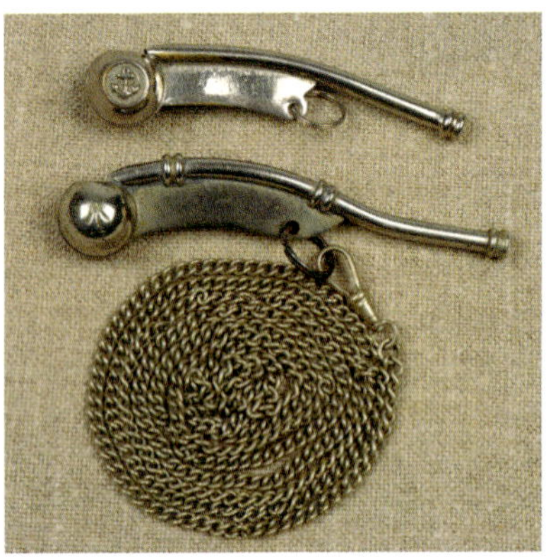

The Boatswain's Call was used by the Quartermaster or Bosun's Mates to make 'pipes'. Pipes would give general orders to the ship's company such as 'pipe down' (lights out) and 'call the hands' (wakey wakey). It was worn suspended by a chain in lieu of a lanyard. *Above* is an early pattern of call, *below* is the standard World War II pattern. From 1937 calls were made available for purchase from Slops.

Admiralty-pattern 'Brushes, Boot, Blacking', and a 'Brush, Hair'. All are marked with the date of manufacture and 'ADMY' for Admiralty.

The standard one-pint tin mug that was available for purchase from Slops. On the mess decks china mugs were usually provided for use at mealtimes.

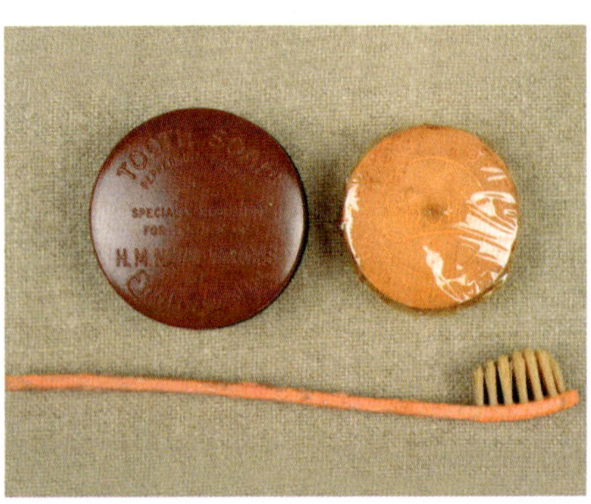

Enolin tooth soap was introduced into Slops in 1936 as a replacement for tooth powder. The block was 3½d complete with case, or 2½d for just the refill at 1944 rates.

An acting PO Quartermaster and two Bosun's Mates pipe the 'side' as HM King George VI boards a Royal Navy warship at Portsmouth during April 1941.

White shoes as worn with the white uniform and tropical clothing. *At left* are the Officers' buckskin shoes and *at right* the Ratings' canvas shoes introduced in 1932. The use of the white canvas shoe was discretionary on foreign stations, but from 1937 one pair was issued to all Ratings proceeding overseas. Both types have plain leather soles and required regular treatment with 'Blanco' to keep them white.

The bridge of escort carrier HMS *Emperor. At left* is the telescope 373B in use. Of interest is the variety of clothing in use, ranging from overalls, to Khaki Drill and White Undress.

At left is a portable signal lamp fitted with a lead for use with a belt-mounted battery pack, for use on beaches or locations were mains power was unavailable. *At right* is a lightweight signal torch.

Representative examples of Royal Navy optics. *Above* is the case for the binocular 1900A and two examples of binoculars, the 1900A with plain lens and with objective lens sun shields. *Below* is the telescope pattern 373B.

A portable signal lamp in use with a Rating from a motor launch. The portable signal lamp was an excellent method of communication at shorter ranges and in good visibility.

On the signals deck of a major warship two ratings communicate using a signal lamp. The rating *at left* is reading return signals with a telescope.

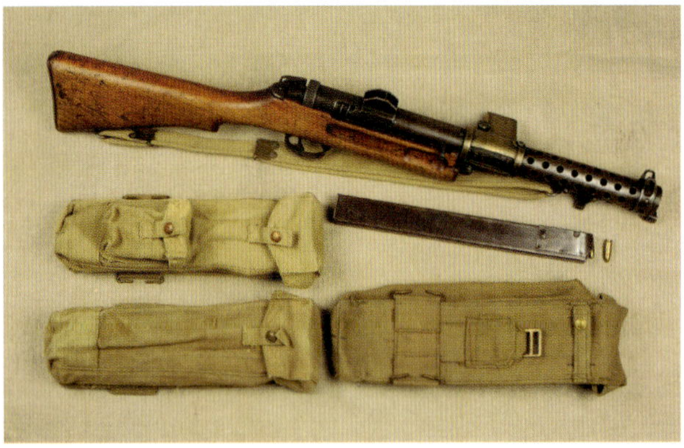

Lanchester Machine Carbine Mk 1*. The Lanchester was a copy of the German MP28 manufactured for issue to the Royal Navy and RAF. Also shown is a fifty-round magazine and pouches.

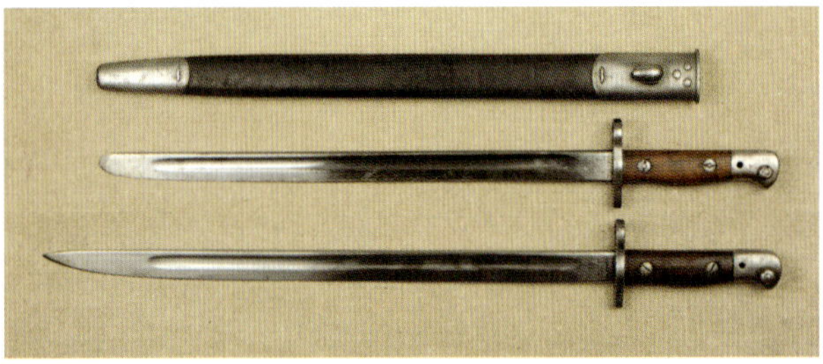

Royal Navy issue '07 pattern bayonets. The '07 fitted both the SMLE and Lanchester. The upper bayonet has a blunted point for use by colour parties.

Seamen kitted out in '08 webbing and armed with SMLE rifles undertake boarding drills. They are covered from the forecastle by a rating with a Vickers gun.

Bayonet fencing practice using an SMLE and '07 bayonet. The rating wears an '08 belt and frog with respirator haversack and Mk II steel helmet.

The White Ensign as flown by all Royal Naval vessels in commission. This small ensign is of the pattern used by minor vessels such as MTBs and motor launches.

The pommel of a Royal Navy issue '07 colour party bayonet showing the stamped 'N' property stamp.

Clothing

Loan Clothing

In addition to the compulsory and optional kit held by Ratings, a number of items were available to them on temporary loan. Loan items included Seaboots, Watch Coats, Duffel Coats, oilskin combination suits and anti-flash. Sun Helmets were a 'permanent loan' item. They were issued to the Rating on loan but remained the property of The Crown in the same manner as seamen's hammocks and kitbags. Such permanent loan clothing could be replaced on a one-for-one basis at no cost to the individual, unlike compulsory kit that had to be replaced at the Rating's expense using kit upkeep allowance.

Dress No. 3A Working Dress for Air Crews

The No. 3A Working Dress for Air Crews was approved for service in March 1945 under AFO 1283. The late appearance of 3A meant that it saw little use during World War II, although BD and BD-style clothing had seen limited use with Ratings aircrew prior to its introduction. Previously Ratings had been required to wear normal working dress which was far from suited to flying duties.

Beret, Working Dress, Air Crews: 3A Dress was worn with a beret fitted with a beret badge or H.M.S. flash, according to rate. For further details of this, *see* the Officer's uniform section under Dress No. 5B Working Dress for Air Crews and Beret, Working Dress, Air Crews.

A non-regulation Battledress blouse badged to a PO Observer. The use of BD by Rating aircrews was not common prior to the introduction of the 3A Dress. Khaki BD was also used as flying clothing.

An Air Gunner 2nd Class 187A badge upgraded to Acting Rating Observer 91A by the addition of spare crown pattern 45A.

RIGHT: The Battledress-style Working Dress for air crews. Numbered as the No. 5B Dress for Officers and No. 3A Dress for Ratings, this blouse is badged to a Leading Air Gunner.

LEFT: Label from the Blouse Working Dress for air crews showing the misleading manufacturer's designation Blouse, Serge, Fleet Air Arm, and manufacturing date of 1945.

RIGHT: The Rating Observer's badge 95A introduced by AFO 4491/42. Similar to that worn by Officers, the badge was to be worn on the cuff of the left sleeve of the jumper or over the left breast on Battledress and the 3A Dress.

Post-July 1944 pattern 96A Telegraphist Air Gunner (TAG) 3rd Class non-substantive badge worn on the Blouse Working Dress for air crews. Prior to the July 1944 reorganization of the airman branch, the 96A badge had been used for 1st, 2nd and 3rd Class TAGs.

Air gunner's wings pattern 96A as introduced under AFO 4491/42. The wings were for wear by Telegraphist Air Gunners 1st, 2nd and 3rd Class. A new series of badges for wear by Rating observers and TAGs was introduced in July 1944.

Beret, Working Dress, Air Crews, introduced in May 1945. At left is the beret worn by other Ratings with H.M.S. flash. At right is the beret as worn by a PO fitted with a Sun Helmet badge. The badge has been modified to a pin clasp fixing.

Suits, Combination, Flight Deck

Introduced in 1942, the Flight Deck Suits were provided for use by aircraft handling parties and other persons employed on the flight deck of carriers during flying operations. The suit was a heavy duty waterproof canvas overall with full front zip, mandarin collar and adjustable cuff and ankle straps. A small flap top pocket was fitted low on the left breast. Issue of the suits was restricted to necessary personnel with an allowance of 350 for fleet carriers, 300 for light fleet carriers and 125 for escort carriers. By 1945 stocks of suits allowed issue to Squadron Commanders, Instructors and Deck Landing Control Officers at Naval Air Stations in the UK, but during winter months only.

Aircraft handlers from the flight-deck crew of HMS *Activity* during 1941. All the men wear the 'Suit, Combination, Flight Deck' with white helmet. Anti-gas goggles are worn as eye protection by five of the men. WWW.ROYALNAVYRESEARCHARCHIVE.ORG.UK

LEFT: Naval airman wearing the Suit, Combination, Flight Deck. This heavy-duty waterproof coverall was issued to all exposed personnel working on carrier flight decks. It is worn with the US M1 steel helmet, anti-flash and respirator haversack. The US M1 helmet was standard kit on ships built and commissioned in the USA.

ABOVE: Flight-deck crew of HMS *Campania* wearing an interesting variety of clothing types during cold weather. The flight deck was particularly exposed to the weather, with wind chill being an additional factor during most deck operations. WWW.ROYALNAVYRESEARCHARCHIVE.ORG.UK

RIGHT: Detail of the Suit, Combination, Flight Deck showing the waistbelt and pocket arrangement.

Helmets, Flight Deck

With all crewmen being dressed alike the identification of Ratings operating on a busy and dangerous flight deck was at best difficult. To assist in the identification of flight deck

The flight-deck crew of HMS *Ravager*. To the front are the DLCO and his team wearing high visibility yellow vests and helmets, behind are the aircraft handlers with their kit, a tractor, starter trolley, chocks and 'starring arms' for aircraft handling. At extreme right are crash and fire crew with asbestos suits and fire extinguisher. WWW.ROYALNAVYRESEARCHARCHIVE.ORG.UK

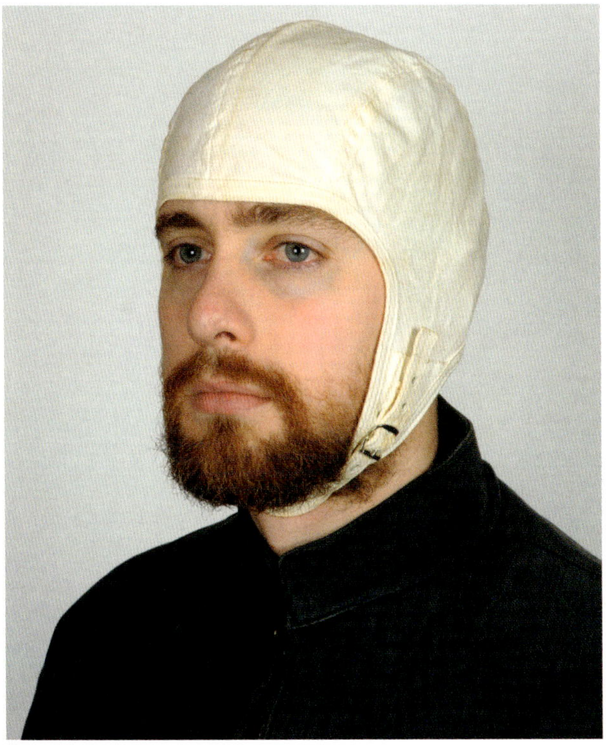

Rather unflattering cloth helmets in various colours were worn by flight-deck crews in order to identify their specific roles. The white helmet was worn by Ratings involved in aircraft handling, fuelling and arming. In 1945 the use of the white helmet was extended to all squadron personnel employed on flight-deck duties.

A DLCO, better known as a batman, wearing the high visibility yellow surcoat, sleeves and cap during operations on board HMS *Battler* during May 1943.

crews, simple coloured cotton helmets were introduced in November 1942. The helmets resembled flying helmets and covered the skull and side of the head, being secured under the chin by a short strap. They were issued in four colours: red for port watch, blue for starboard watch, white for Ratings involved in aircraft handling, fuelling and arming, and yellow for aircraft directing. In addition to helmets, yellow sleeves and jackets also provided high visibility for those involved in aircraft directing. In 1943 the blue helmet was replaced by the more obvious green for starboard watch personnel.

In addition to the four basic colours, identification numbers or distinguishing marks could be painted on the helmets to identify particular sub-duties if required, such as differentiating between handlers and armourers.

In October 1945 the flight deck helmet colours were amended as follows: red, firemen; green, special duty men; white, squadron personnel employed on the flight deck; yellow (also sleeves and jackets), flight deck Officers and aircraft directors; blue, general duties; halved black/white, flight deck engineer party. Team numbers of members of aircraft handling parties were also in future to be painted on the rear of the helmets in 2in figures.

Watch Coat (Ratings)

Watch Coats were a loan clothing item issued to men who, by the nature of their duties, were exposed to inclement weather, including such duties as watch-keepers on board ship and sentries at Royal Naval Barracks. Issue levels were heavily restricted, and typical rates were thirty-two Watch Coats for battleships, carriers and heavy cruisers, twenty-four for light cruisers and destroyers, and eight for minesweepers. Issue levels for shore establishments were typically provided for, at one coat per man on duty at any given time. AFO 77/1945 introduced a new, smaller size for Watch Coats, size 0 to fit chest sizes 32 to 35in that was suitable for both WRNS and smaller naval Ratings.

The Watch Coat was a single-breasted wool Great Coat with five buttons at the front, and flap top pockets at each hip. It had a large fall collar that could be worn up for added protection.

RIGHT: RN-pattern web leggings. This pair was manufactured in 1939. They have four laces and a single strap and buckle fastening. These replaced an earlier pattern of legging that used a second strap instead of the lowermost lace. Prior to 1921 leggings had been made from flax (linen).

A group of general service Ratings pose for the camera. The sailor at right wears the Watch Coat and 1908 webbing. The respirator haversack is worn at the 'alert' position on his chest.

LEFT: The single-breasted Watch Coat worn by a naval sentry. The sentry is equipped with steel helmet, 08 web equipment, SMLE rifle, and a whistle on a chain to provide a means of raising alert.

Duffel, Long, Coat

The wool Duffel Coat had been in use in the 1890s, having a three toggle fastening and a plain hood with no adjustment. Early World War II Duffel Coats had a three-toggle fastening, adjustable wrist straps, neck flap and press snaps to adjust the hood to fit. A large open-top pocket was fitted at each hip. During the war economies were introduced, the neck flap and snaps were both deleted, and the hood adjustment was effected by a tape that ran around the inner edge of the hood and which could be drawn tight and secured with buttons. In addition the front was given four toggles, improving the closure.

Duffel Coats were intended for use by men at exposed positions, such as armaments, control and searchlight positions. They were not intended for ordinary navigational work or watch keeping, for which the Watch Coat was to be used. Typical scale of issue was from 200 to 250 for battleships, to twenty-five for destroyers and eight to ten for submarines. Small vessels such as MGBs and MTBs were not authorized an issue of the Duffel Coat. The coats were issued for the period of duty only.

The early production Duffel Coat worn by a naval Officer on the bridge of an HDML. The early duffel was fitted with a throat tab, hood fastening snaps and three-toggle front fastening.

ABOVE: The early production Duffel Coat worn by sailors engaged in an impromptu sing-song on the forecastle of a destroyer.

The Duffel Coat was provided in only three sizes – being oversized to fit over other warm clothing, it was unnecessary to have a perfect fit. Issue of the coat was discontinued in 1956.

RIGHT: Officers on the bridge of HMS *Starling* wearing a mix of clothing types. The two Officers at left wear the early war Duffel Coat, while Commander Walker (centre, taking a compass bearing) wears the blue Royal Canadian Navy version of the duffel occasionally seen in use with the RN.

FAR LEFT: The early production Duffel Coat hood had brass snap fasteners at each edge to allow the face aperture to be adjusted and closed down in extreme weather. A neck flap was also fitted across the front of the throat.

LEFT: Visual signalman. This 'Bunting' wears the later variant duffel with four-toggle fastening and a draw tape and button adjustment for the hood.

A submariner displays fresh fruit and vegetables, a rare treat for submariners. He wears a typically grubby white frock as associated with the submarine service. The wartime frock had a distinctly short collar.

Frock, White (Submarine Sweater)

The warm woollen frock issued to submariners was in use during the Great War; interestingly an almost identical garment was also used by German submariners. Two of these expensive garments were issued to each Rating and Officer of the submarine service, as part of the submarine protective clothing, and to coastal forces crews. The frock was also provided for use by boys serving at the training establishments HMS *Ganges*, *Caledonia*, *Wildfire* and *Impregnable* for winter wear when undergoing boat training, flight deck parties on aircraft carriers of the China and home fleets, and to catapult-crews in ships of the home fleet.

Typical of period documentation, where written material was produced manually and on a worldwide scale, the frock is listed in a number of guises, depending on the document source: thus Frock, White; Sweater, White; Submarine Sweater; Submarine Frock; and Frock, Woollen, Submarine Service, all refer to the same garment. They were available in three sizes: 36, 38 and 40in.

During the 1930s the Admiralty received numerous requests for the frock to be made available to the Fleet for purchase from Slops. The request was not granted.

Oilskins

Oilskin suits of jacket, trousers and Sou'westers were provided as loan clothing for use by boat crews, men employed on anchor work, and by men otherwise subject to exceptional exposure during the course of their work. These were loan clothing items held in limited quantities to be issued as needed. Battleships, aircraft carriers and cruisers would hold 100 suits, with smaller ships being allocated fewer, depending on their complement. Typically six to eight suits were provided for submarines, and one per man for MTBs and MGBs; these were later replaced by specialist clothing, the Waterproof Protective Suit for submarines, and the Weatherproof Protective Suit for MTBs and MGBs.

Prior to 1932 the oilskin coat had been a double-breasted garment, but in 1932 a new pattern of oilskin coat was introduced. It was single-breasted, had a narrow skirt with a vented back slit, and storm cuffs and pockets.

Class II Ratings were issued an oilskin coat as compulsory clothing. A pair of oilskin leggings was an optional holding. Oilskins needed special storage conditions if they were not to become tacky and succumb to excessive deterioration. They had to be stored hung up in a well ventilated, cool and dry area away from heat sources. Oilskin was at risk of spontaneous combustion if left in piles in unventilated areas. They were especially prone to severe deterioration in tropical regions and Class II Ratings proceeding to the East Indies were not required to hold an oilskin in their compulsory kit, and were allowed to sell them prior to drafting.

This gunnery Rating wears a fire-control communication tube headset with oilskin coat and sou'wester. The oilskins provided excellent waterproofing but offered little in the way of insulation and were prone to a build-up of perspiration, leaving clothing damp and further reducing insulation.

Waterproof Protective Suits (Submarine or Ursula Suits)

The two-piece Waterproof Protective Suit, or Ursula Suit as it is better known, consisted of a waxed cotton jacket and trousers. The jacket had an integral hood, a drawstring waist, adjustable cuffs and two expanding bellows pockets at the hip. The trousers had reinforced knee sections and integral braces. The suit was officially introduced into the RN under AFO 1588 of April 1942.

The suit's name came from the submarine *Ursula*, commanded by Lieutenant Commander George Phillips, who is credited with designing the suit. Philips wore a waxed cotton suit, produced by the Barbour company to his own specification, on patrol. The waterproof ensemble was a great success, and in a modified form was adopted by the Admiralty for issue to submariners.

AFO 4310/44 listed four sizes of the suit. Issue of the Ursula suit was very restricted. Submarines were provided with one suit for each executive Officer, five for lookouts and one for the signalman. Two pairs of mittens were supplied with each suit. Reserve stocks were correspondingly low, with submarine depot ships carrying a stock of five suits for each submarine attached.

The crew of HM Submarine *Ursula* gather on the boat's casing in December 1939. Lt Lakin and Lt Cdr Phillips at the front of the group wear waxed cotton jackets, produced by the Barbour company for Phillips, and which inspired the Waterproof Protective Suits better known as the Ursula suit. Lieutenant Commander George Phillips is the bearded officer with binoculars.

The Destroyer Suit in use with the crew of a quadruple .50in AA gun. This view shows the snaps fitted to the lower outer leg seam, allowing the garment to be donned or doffed over footwear.

LEFT: The Weatherproof Protective Suit, better known as the Destroyer Suit, worn by men of a Coastal Forces craft. The one-piece oilskin suit had two large thigh pockets, adjustable waistbelt and a warm kapok liner.

Weatherproof Protective Suits (Destroyer Suits)

It had been found that the available cold weather clothing, typically duffel coats and oilskins, was inadequate for those working on the upper deck of destroyers in cold climates. A special weatherproof suit was designed and manufactured for men on destroyers serving in Home and North Atlantic waters. The suit consisted of a one-piece oilskin garment with detachable hood and a detachable kapok liner, and was available in three sizes; it was brought into service under AFO 4538 of October 1941. The suits were a welcome addition to the equipment of personnel on destroyers, and soon became known as the 'destroyer suit'. As production met demand the suits were issued to other small craft crews, including those of minesweepers, MGBs and MTBs. It was found that some men wore the detachable liner without the oilskin outer. The suit was expensive to produce and used valuable resources so such misuse was actively discouraged. The standard scale of issue was fifty suits per destroyer.

The kapok lining led some to believe that the suit was buoyant, as kapok was used in lifejackets of the period. However, AFO 2200/43 advised that the quality and quantity of kapok in the suit was suitable only for insulation and not for floatation. When wearing the weatherproof suit the inflatable life belt was also to be used.

From 1943 a special waterproof cape was produced for use with the destroyer suit. The capes were provided only for use by gunlayers manning power-operated mountings of Coastal Forces craft.

MTB Suit

Coastal Forces crews were issued with the Weatherproof Protective Suit. However, it was felt that the suit was inadequate for the conditions under which Coastal Forces operated, so during the latter part of 1944 a special suit for MTB crews was developed by GQ Parachutes. The suit had heavy-duty zips, was lined with high grade kapok padding, which gave it good buoyancy, and it had an integral inflatable 'Mae West'. For additional warmth a kapok-lined inner jacket was used. It was an expensive item but it was found to be superior to both the Destroyer Suit and the Submarine Suit in every way, including heat retention, waterproofing and flotation properties. In tests the suit was found to provide sufficient buoyancy even when worn with Coastal Forces body armour. The suit saw limited operational testing in the closing stages of the conflict against Germany, and by November 1945 sufficient MTB suits had been produced to equip all upper-deck crews of Coastal Forces craft.

Overalls

These included blue, two-piece overall suits, two-piece overalls, and Combination (Blue Overall) Suits. During the Great War and the inter-war years Ratings had been issued a two-piece Canvas Overall Suit. The two-piece suit had a single-breasted jacket with stand collar and four-button front. The trousers had a fly front and an integral belt and buckle. The suit was not issued to Officers' stewards or Officers' cooks, who had no requirement for dirty work clothing. Stokers were not issued the suit, but were provided with Fearnought trousers, canvas jacket and flannels.

Artificers wore a Blue Combination Suit instead of the Canvas Overall Suit. In 1930 AFO 1838 introduced a new pattern of two-piece Blue Overall Suit with bib-and-brace trousers for Artisans, as an optional item of

clothing in lieu of the Combination Suit. The use of the Combination Suit and Blue Overall Suit was later extended to other Ratings, replacing the Canvas Overall Suit. The two-piece Blue Overall Suit was an optional item for use by all Ratings except engine room Ratings. In 1940 a new pattern of trousers was issued with beckets (loops) for wear with a belt. The trousers allowed the jacket to be tucked inside the waist when working with machinery, an important safety consideration, and were cooler than the bib-and-brace in a hot environment.

The Combination Suit had a button front closure, a button top pocket on the front left breast, and a side pocket at each hip. In 1939 AFO 1192 introduced a modified pattern of one-piece Blue Combination Suit with a fly front and plain top pocket.

ABOVE LEFT: The two-piece Blue Overall Suit of the pattern introduced in 1937. It consisted of plain blue trousers and a simple jacket with two skirt pockets and a four-button front fastening. A small open top pocket was fitted to the inner right face of the jacket.

ABOVE RIGHT: A Submariner wearing the Blue Combination Suit with fly front, introduced in 1939. They are worn with the blue waistbelt fitted with a spring clip for the suspension of the knife. The owner's ID tag is fitted to the suspension ring.

Crewmen from HM/SM *Thunderbolt* show a mix of work clothing. The 'Jimmy', Lieutenant J. Stevens, wears the Undress Coat, while to his right a submariner wears the two-piece Blue Overall Suit.

ABOVE: A stoker and Chief stoker wearing single-piece overalls in a warship's engine room. The stoker manning the telegraph repeater wears the early pattern Combination Suit with visible buttons. At right the Chief stoker wears the post-1939 fly-fronted pattern.

Ratings from the battleship HMS *Repulse* wearing the single-piece Blue Combination Suit with fly front. The sailor third from right wears the trousers for the two-piece Overall Suit.

An illustration of the Action Working Dress produced for publication to advise Officers and Ratings of the new uniform. The AWD was listed as No. 14 Dress for Officers and No. 8 Dress for Ratings. Despite the illustration showing pocket flaps, early production shirts had open top pockets.

Airmen from 1820 NAS at US Naval Air Station Brunswick during 1944. The majority of the men wear US Navy dungarees and chambray shirts as working rig, a common practice for units working in the USA and during the Pacific campaign.

Action Working Dress No. 8 Dress worn by trainee Ratings in the immediate post-war period. AWD saw very limited issue prior to the end of hostilities. The Ratings at left and right wear the austerity-pattern blue waistbelts.

Action Working Dress (AWD; Action Dress)

One of the most innovative advances in naval uniform came about with the introduction of the Action Working Dress. Officers' AWD was introduced by AFO 2125/45 as '*Uniform – Officers, Action Dress*' and was to be No. 14 Dress. Ratings' AWD was introduced by the next AFO in sequence, AFO 2126/45, and was to be No. 8 Dress. AWD could be worn by Ratings at all times that the No. 5 Dress had previously been used. It was anticipated that No. 8s would replace No. 5 Dress in peacetime once the use of No. 6 Dress was resumed on cessation of hostilities. The AFO stated that patterns had been sealed in a substitute fabric due to shortages of the preferred type, and that a new pattern would be introduced as the better fabric became available. Owing to production difficulties it was anticipated that there would be a considerable delay before stocks were available for issue. Priority of issue was to be for aircrew and Ratings serving afloat in the East Indies and Pacific stations.

AWD was to be worn as an action and anti-malarial dress in the tropics, or as conditions required. In conjunction with the anti-flash, and with trousers tucked into socks, it was designed to provide maximum protection from explosive flash and fire; it also protected the wearer from mosquito and insect bites, and was primarily to be worn when at action stations.

A two-piece suit, it comprised a long-sleeved, light blue cotton collar attached shirt with six-button front. A plain open-top pocket with button closure was fitted at each side of the breast. The fly front trousers were of dark blue cotton with two adjustment buckles to the front and beckets for the fitting of a belt. There were two side pockets and a hip pocket. It was to be worn with the black cap or cap with white cover as ordered. The rig was not dissimilar to the US Navy's dungarees and chambray working dress, which was frequently seen in use with the RN in the Pacific. Officers were able to purchase No. 14 Dress from service sources but at their own expense. AWD was not issued in any quantity before the cessation of hostilities, but it was to become the standard working dress of the post-war Navy.

Equipment

Lifebelts

Cork lifebelts were standard well into the early part of World War II. Made from a heavy canvas jacket fastened with a waist tie, the front and back sections each contained four large cork blocks over the upper chest and back, and three smaller blocks at the waist and lower back held within the canvas cover.

In October 1939 AFO 3200 announced that a new inflatable lifebelt had been introduced and was to be provided as personal issue to all Officers and Ratings of ships' complements (an inflatable 'swimming belt' was issued during World War I, but was discontinued in 1921). As well as being manufactured in the UK, arrangements were made to have lifebelts produced in Australia for issue to the China and East Indies Fleets. By the end of 1939 many ships had already received their full allowance. However, the speed of production led to many lifebelts passing through the system that were later found to have life-threatening defects.

A confidential order issued in December 1939 advised that the inflatable lifebelt should not be inflated until the wearer was in the water as there was a risk of getting fouled in the superstructure and being sucked down with the ship. The order was revised three months later, instructing instead that all lifebelts should be inflated before entering the water, unless there was a danger of being trapped in the superstructure.

By early 1940 all men at sea had been provided with an inflatable lifebelt. Production of lifebelts was maintained so that a personal issue could be made to all Officers and Ratings serving ashore, except for those who were unlikely to go to sea. However, the cork lifejacket was still retained for some heavy evolutions, including use by sea boat crews and gun crews, but it eventually fell into disuse in favour of the inflatable pattern.

The inflatable lifebelt was a simple device consisting of a rubberized tube with a protective blue wool knit or blue duck sleeve. The sleeve had a long loop of tape that passed behind the head and supported the lifejacket around the neck, adjusted so that the lifebelt was correctly positioned high on the chest. Side straps fitted directly to the inflatable tube secured the lifebelt to the body. An oral inflation tube was positioned centrally on the inflatable tube. The inflatable lifebelt was also produced with a white duck cover.

Use of the inflatable lifebelt under wartime conditions highlighted issues with the method of wear. If worn too low it caused problems with the floating position when in the water, often tipping the wearer's upper body into the sea due to the low centre of buoyancy. If tied incorrectly the lifebelt also had a tendency to pull away from the wearer, reducing the support and again altering the floating position in the water. The lifebelt was correctly worn with the upper edge at breast level and with the tapes crossed at the back and then knotted above the belt at the front. This manner of wear helped prevent the lifebelt coming away from the body and kept the centre of buoyancy high.

In early 1942 combat reports from survivors of sunken ships indicated that unless lifebelts were actually being worn there was little opportunity of donning them when emergencies arose. Inflatable lifebelts could be worn at all times with little discomfort or interference with performance of duties, so it was ordered that they were to be kept on the person at all times when the risk of damage from enemy action was high.

By 1943 'life-saving torches' and sleeves were available for use with the lifebelt. Manufactured by Easco, the torch consisted of a green metal battery case with screw-top lid, cable, lamp and attachment clip. The clip allowed the lamp to be attached to the clothing or lifebelt while the battery case was stowed in a fabric sleeve-type pocket that fitted to the lifebelt.

During the early stages of World War II the cork lifebelt was the standard personal buoyancy aid. It consisted of cork blocks fitted into a canvas jacket. It was a bulky item that restricted the movement of the wearer. NATIONAL WAR MUSEUM MALTA

A sea-boat crew recover a torpedo from the water. The men wear the cork lifebelt. Following the introduction of the inflatable lifebelt the cork lifebelt remained in service for use by boat crews and tasks where the inflatable belt would be subjected to undue wear.

The standard Royal Navy inflatable lifebelt with its blue duck cover. Introduced in 1939, this item was personal issue to all Ratings and was to be worn at all times when the risk of damage from enemy action was high, including when off watch and sleeping. The belt was manually inflated using the oral inflation tube.

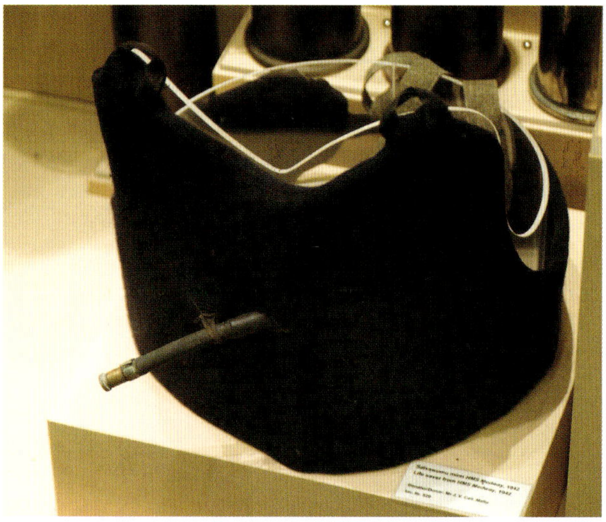

This rare early variant of the inflatable lifebelt had an abdominal section and a wool-knit cover. This pattern was predominantly used in the Mediterranean. NATIONAL WAR MUSEUM MALTA

LEFT: This lieutenant wears a variant of the inflatable lifebelt with white duck cover. Of note are the helmet with rank marking on the side and the anti-flash hood and mask.

RIGHT: A smiling Rating serving on a coastal minesweeper wears the inflatable lifebelt over his white frock.

A gun crew wearing the inflatable lifebelt with abdominal section come to action somewhere in the Mediterranean. A mix of Mk I and Mk II helmets are worn.

Walcheren, November 1944. The RN and RM crew of a landing-craft gun (LCG) abandon ship as the vessel sinks following a hit by German shore batteries. The men are wearing the inflatable lifebelt.

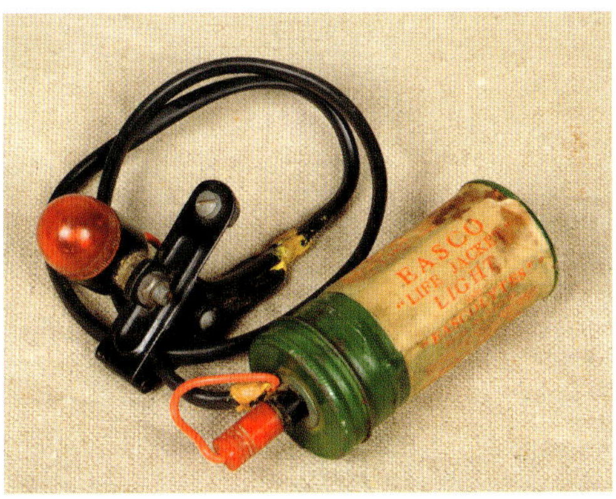

The inflatable lifebelt was provided with a 'life-saving torch'. Issue of the Easco life-saving torch began in 1943.

RIGHT: Various floatation devices were available from naval outfitters. One such item consisted of the Officer's Morning Waistcoat fitted with an internal rubber stole that could be inflated via a tube through the breast pocket. They were popular with RN as well as Merchant Officers, although their use was discouraged in favour of the issue lifebelt.

Body Armour (Flak Suit)

Towards the end of World War II a quantity of Wilkinson Grow flak suits were procured by the Royal Navy for issue to Coastal Forces. The flak suits were Armor, Flyer's, Vest M1, as issued to the USAAF and RAF aircrew. The vests, marked with a large 'N' property stamp, were held at Priddy's Hard Royal Naval Armaments Depot for issue to Coastal Forces units as required. Post-war figures show that as many as 1,200 of the vests were held in store.

In July 1945 the Commanding Officer of HMS *Formidable* wrote to the Flag Officer Commanding First Aircraft Carrier Squadron, British Pacific Fleet, making a request for the issue of American pattern flak suits to personnel whose action stations left them exposed and unprotected. The request was given some consideration and eventually reached the Secretary of the Admiralty in late August 1945. However, by the time it arrived the war had ended, but it was suggested that although an immediate requirement for armour was no longer necessary, consideration should be given for the permanent introduction of flak suits into the Royal Navy.

The Wilkinson Grow-pattern 'flak suit' was issued to Coastal Forces in 1945. The armour was of the USAAF aircrew pattern Armor, Flyer's, Vest M1. The armour is worn with the RAC Mk I helmet.

Boots, Sea (Seaboot)

Leather Seaboots, introduced in the mid-nineteenth century, were issued to men whose duties left them exposed to the weather during upper-deck evolutions in cold climates. AFO 958/37 introduced rubber Seaboots into service, with future holdings of Seaboots to be 50 per cent rubber, 50 per cent leather. A supply of rubber Seaboots was also held by most vessels for use by decontamination parties. Seaboots were issued oversized so they could be worn with thick Seaboot stockings and a pair of lighter stockings for insulation. The oversize boot and thick stockings allowed easier removal of the boots, an essential attribute if the wearer found themselves in the water.

Seaboots were tested for waterproofing by filling them with water to a depth of 9in: if the boot leaked it was rejected. An AFO issued in September 1944 highlighted the

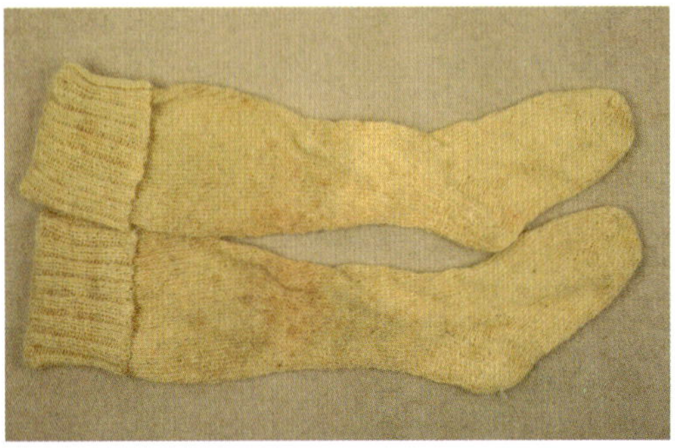

LEFT: Admiralty-issue Seaboot stockings. These heavy wool stockings were popular with men working on the upper deck and were worn as an insulating layer with the oversized rubber or leather Seaboots.

The heavy duty Seaboot used by Ratings working on the upper deck and exposed to the elements. The boots were waterproof leather and were made oversized allowing heavy wool Seaboot stockings to be worn for insulation. The collar of the boot had a distinct curve to the forward edge.

LEFT: The standard black leather lace-up 'Boots, Half'. The leather soles were plain and unstudded. All Class I, II and III Ratings were required to have two pairs of half boots or one pair of boots and one pair of shoes. Under training, men were required to stud the soles of one pair of boots to reduce wear, but the studs had to be removed before they were drafted to sea.

A seaman gunner manning the loading tray of a high-angle AA gun on the battleship HMS *Courageous* during March 1937. Leather Seaboots are worn with the boot collar front turned down.

Rubber Seaboots being worn by a gunner manning a .5in AA gun. The boots are worn with Seaboot stockings with the top turned over the boot collar, a common practice that helped prevent the stockings slipping down during wear.

lack of care being afforded to leather Seaboots. They were often returned to stores unpaired and worn down to a level that required scrapping. Repair of minor stitching and the replacement of soles and heels could only be effected if the boots were returned to stores before they had deteriorated beyond repair. The order emphasized the use of dubbin or other suitable grease to preserve leather, and pointed out that the boots were not only costly but their production required specialist skilled labour and valuable raw materials that were already scarce.

Anti-Flash

Anti-flash was in use during the Great War. It was issued to protect exposed personnel from high explosive flash and fire. In peacetime, anti-flash was issued at a rate of 4 per cent of ship's crew, with a minimum of eight sets of anti-flash helmets and anti-flash gloves being held in store for issue as required. At the outbreak of war, anti-flash was to be provided in sufficient quantity to equip the whole ship's company with a set on personal loan, plus an allowance for replacement. In smaller vessels, such as trawlers, it was not felt necessary to provide anti-flash to those other than gun crews.

The fireproof finish of anti-flash was destroyed by washing or extensive soaking in water. Its effectiveness also declined over time and with use. Thus all anti-flash that had been in use for twelve months or had been washed or otherwise soaked in water had to be re-proofed. This was achieved by soaking the anti-flash in a mixture of crystalline borax, boric acid and water.

Anti-flash consisted of long-sleeved gloves and an open-faced balaclava-style helmet. However, the reality of warfare at sea soon found the anti-flash helmets provided insufficient protection, as they left the face exposed. In order to improve the effectiveness of the ensemble an

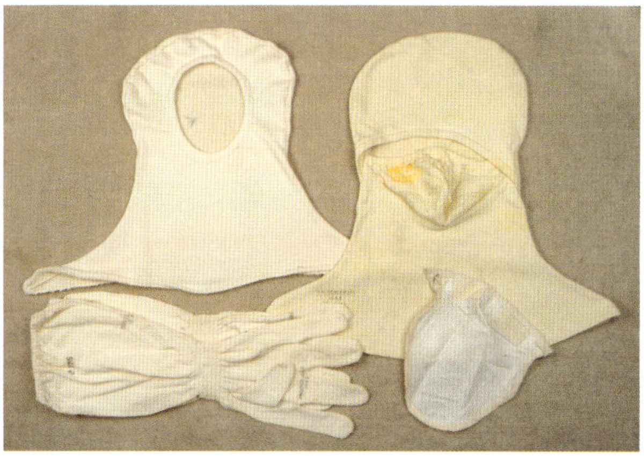

Anti-flash. At top left: the anti-flash helmet in use since the Great War. Top right is the anti-flash hood with integral face protection, introduced in 1943. The anti-flash helmet had an open face, and in 1941 it was enhanced by the issue of a mask, shown lower right. At bottom left are anti-flash gloves.

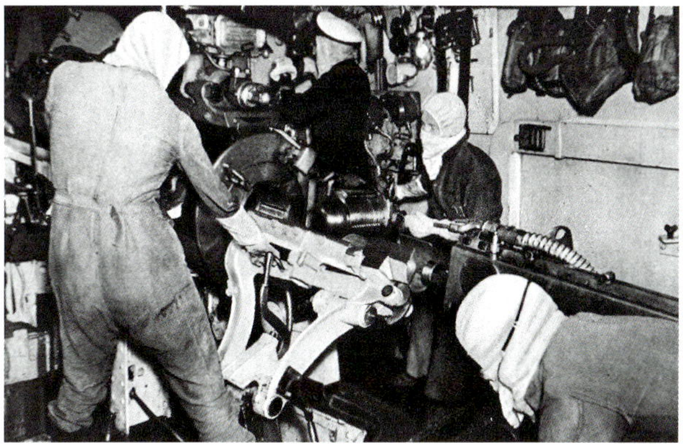

The anti-flash helmet worn with the anti-flash mask by a heavy gun turret crew. Of note are the respirators hung on the bulkhead at top right.

A 40mm Bofors AA gun crew at action stations on board the escort carrier HMS *Emperor*. The men wear anti-flash helmets and overalls. WWW.ROYALNAVYRESEARCHARCHIVE.ORG.UK

LEFT: The early war anti-flash helmet worn over the anti-flash mask. This method of wear provided good protection but the mask was commonly worn over the helmet, greatly reducing its effectiveness as it would be lifted away by any blast. GUN COURTESY EXPLOSION – MUSEUM OF NAVAL FIREPOWER

RIGHT: The anti-flash hood with integral face mask introduced in 1943. GUN COURTESY EXPLOSION – MUSEUM OF NAVAL FIREPOWER

anti-flash mask was introduced in 1941 to cover the nose and mouth area. The cellular cotton mask was held loosely over the anti-flash helmet by a single adjustable elastic strap. The mask was not a perfect solution as it could be lifted away from the face by blast. The solution was to have an anti-flash helmet with an integral mask. The new anti-flash hood with integral face protection was available from 1943.

During 1943 trials of a lightweight anti-flash hood were instigated. Thirty hoods were manufactured and hot weather trials conducted. The trials were inconclusive, but it was generally agreed that the lightweight hood was more comfortable than the standard pattern despite the lower level of protection offered. A further batch of hoods was tested during 1944 but no further progress was made with lightweight anti-flash during the war.

It was emphasized that for full flash protection all exposed skin should be covered, sleeves were to be worn rolled down, and trousers tucked into the top of socks. Anti-gas goggles could also be worn to provide protection for the eyes, but their use was not recommended in tropical regions as they were prone to misting up. Tropical clothing was not to be worn in action as it left extensive areas of the skin exposed. However, this ruling was all too frequently ignored.

Helmet, Steel

In January 1930 AFO 160 ordered that in order to make them less conspicuous, all steel helmets held by the Navy were to be repainted and sanded. This required that the surface be sanded to bare metal and then painted with dark grey paint. While the paint was still tacky a layer of sand was to be sprinkled over the entire surface. After twenty-four hours the surface was coated with thinned paint to bond the finish. At this time the holdings of steel helmets, originally termed 'shrapnel helmets', were held solely for use by landing parties ashore.

The standard Mk II helmet was provided for all upper-deck crew exposed to enemy fire. The example *at left* has been personalized by an RN Commander. Officers often obtained personal helmets, which were marked with rank to make them readily identifiable when wearing anti-flash. *At right* is the RAC Mk I helmet issued in limited numbers to Coastal Forces from October 1943.

Gunnery quarters Ratings man a 4.7in gun. Anti-flash helmets and gloves are worn, with an anti-flash mask also being used by the Rating at left operating the breech opening lever.
GUN COURTESY EXPLOSION – MUSEUM OF NAVAL FIREPOWER

ABOVE: AA fire-control Ratings scan the sky for enemy aircraft. The men all wear the Mk II helmet, at least two of which have been personalized with markings to the front.

RIGHT: 40mm Bofors AA crews on board HMS *Battler*. While the communications Rating in the foreground wears the standard British Mk II helmet, the majority of the gunners wear the US M1 helmet.

The 'protective eye vizor' was issued to certain RN and RM personnel as protection against flying splinters. 10,000 helmets with fitted vizors were issued for trials in October 1940: 3,200 to the Navy, 6,000 to the Royal Marines and 800 held in reserve.

Royal Navy-issue Mallock Armstrong ear defenders. These were introduced in 1915 to help reduce the instances of deafness due to gunfire.

In the spring of 1939, war with Germany seemed imminent. In order to prepare for conflict, warships of the Home and Mediterranean stations had approved scales of establishment of 'Helmets, Steel' drawn up for ships. Small quantities of helmets were already held, but new scales allowed for increased numbers and added provision for all ships for use by landing parties and exposed personnel.

In the spring of 1940 a special eye visor, providing protection from low velocity fragments, was produced for limited issue to certain RN and RM units. The visors were fitted to the steel helmet and weighed 3½oz. An initial supply of 10,000 helmets with visors was procured, with 6,000 going to the Royal Marines, 3,200 to various naval commands, and the remaining 800 held in reserve. The helmets with visors were delivered in October 1940 for trials. It was a requirement that reports as to the desirability of adopting the visors, or otherwise, be supplied to C-in-C Home Fleet. Following extensive trials the helmets with visors were not adopted as standard for fleet use.

It was felt that the Mark II helmet on general issue to the Royal Navy did not best meet the special needs of Coastal Forces. In October 1943 it was decided to introduce the RAC Mk I helmet for use by Officers and upper deck Ratings of Coastal Forces craft. Priority for the issue of the limited stocks was to go to gunlayers on selected craft.

The crew manning this quadruple .5in AA gun all wear the 'blast helmet'. The blast helmet, introduced in 1940, was similar to a scrum cap and provided excellent noise protection to the ears, far superior to the Mallock Armstrong ear defenders. The helmet was designed to be worn with a steel helmet, but it was difficult to don anti-flash over the padded side cups.

Ratings' Sun Helmet, post-1936 pattern bearing the cap ribbon of H.M.S. FROBISHER. This pattern of helmet was used until April 1944, after which time the use of Sun Helmets was discontinued in the RN This pattern of helmet with rounded rear brim was also occasionally used by Officers, with or without the seven-fold pagri.

A 1932-dated image of seamen manning a whaler. The Sun Helmet is worn with a cap ribbon by all the seamen.

Naval Sun Helmet (Ratings)

The Sun Helmet had been introduced in 1921 to replace the Sennit Hat under AFO 664. The helmet had a white drill outer surface and a distinctly oval shape, lacking the peak and broad rear of the Wolseley helmet. Prior to 1921 CPOs had worn a Sun Helmet of the pattern supplied to Royal Marines, but with the introduction of the new helmet CPOs were required to wear the new pattern. When issue of the new helmet was announced, orders allowed ships on foreign stations to purchase helmets from local sources pending the issue of the approved pattern. This led to some variety in the helmets used during the early years of the 1920s until the standard patterns had reached the Fleet. In December 1921 special cap ribbon retaining beckets were adopted for use with the helmet. Prior to 1939, and the introduction of special badges, Class I and III Ratings wore the helmet with no pagri or badges.

This CPO demonstrates the correct method of presenting a salute, with the upper arm parallel to the shoulder and the index finger touching the forehead just above the eyebrow. He wears the White Drill Suit with web leggings and Sun Helmet. BR159 1934

The Rating's pattern Sun Helmet was worn by all CPOs, POs and men serving on foreign stations as required. As a result of the Review of Service Conditions, held in 1936, where the style of Sun Helmet was the subject of frequent complaint, the Admiralty approved the introduction of a slightly modified pattern of helmet that was considered smarter and provided better protection to the back of the neck. For comfort the helmets were usually issued one size larger than the wearer's normal cap size.

A small stock of Sun Helmets was maintained for loan to Royal Marines working on the upper deck to protect them from the sun. In addition a supply of old helmets unfit for normal wear was held for loan to men working over the ship's side, in painting parties or similar employment.

Ratings working on aircraft at RN Air Stations wearing khaki tropical clothing were provided with a khaki Sun Helmet on loan.

In 1943 AFO 3214 discontinued the use of cap ribbons on Sun Helmets for the duration of the war. In April 1944 the issue of Sun Helmets, covers and Sun Helmet badges ceased, and all helmets, calico covers and badges held by Ratings were withdrawn. However, in order to meet any special requirements for use by sentries, landing parties and similar activities, small stocks were to be retained by ships and establishments.

The tropical sun helmet worn by a CPO. The sun helmet could be worn in lieu of the cloth cap with white cover with No. 5, 6, 7 and 10 Dresses.

Ratings' Sun Helmet of the improved pattern introduced following the Uniform review of 1936. It provided better protection to the back of the neck than the previous 1921 pattern with its oval brim. The helmet bears the CPO's pattern 53 badge.

Metal badges introduced in 1939 for wear on the Sun Helmet. *Left to right:* blue-painted pattern 55 badge for men not dressed as seamen, bi-metal pattern 54 badge for confirmed POs, and the bi-metal pattern 53 badge worn by CPOs. Men dressed as seamen wore a cap ribbon on the Sun Helmet.

Rating's Sun Helmet Badges

The requirement for identifying badges to be worn on the Sun Helmet by CPOs, POs and men not dressed as seamen had been identified and discussed at length during 1937. It had been acknowledged that in tropical dress with Sun Helmets individual ratings were not evident and that a badge was needed for wear on the Sun Helmet. Considerable effort went into the planning of the design of badges. Lugs had been favoured initially but required holes to be made in the helmets, so prongs were decided upon. By the spring of 1938 patterns had been sealed, in October contract CP 42981/38 was placed with Messrs Wm. White (Birmingham) Ltd, and by early 1939 production had commenced.

In May 1939 AFO 1266/39 introduced the metal badges for wear on the Sun Helmet by Ratings in Class I and III uniform. Class II Ratings were required to wear the cap ribbon. The new badges were a simplified metal version of the embroidered cap badge: a silver anchor with brass rope and laurel wreath surmounted by a crown for CPOs (pattern 53), a silver anchor with brass double rope surround surmounted by a crown for confirmed POs (pattern 54), and an anchor with double rope surround and laurel wreath surmounted by a crown, all in brass and painted in blue enamel for men not dressed as seamen (pattern 55). Although the cap badge usually worn by Class III Ratings was red, it was felt that for wear with the Sun Helmet a blue enamel badge would match the blue badges worn with white clothing. Post war the CPOs pattern 53 and POs pattern 54 badges were re-designated beret badges.

Knife, Clasp (Knife, with Spike)

The clasp knife was compulsory kit for seamen branch Ratings in Class II uniform, and was to be worn at all times when in working rig. It was an optional item of kit for other classes of Rating.

In 1929 the horn-gripped knife with blade and marlin spike was replaced by a new pattern with slab-sided metal grips. The new knife was the subject of a series of complaints of blade breakage, the majority of which were not upheld as any shortcomings were generally found to be due to misuse. In 1937 a slightly modified knife was introduced. Slightly smaller than the 1929 pattern, the new knife incorporated a screwdriver. The clasp knife had a blade on one side and a marlin spike for ropework on the opposite side. The slab sides were deeply cross-hatched for grip and had a plain flat panel on which the owner could stamp his name.

The knife was worn on a clip suspended from the waist belt or on a lanyard.

LEFT: It was common practice for the clasp knife to be suspended from the waistbelt using a carbine clip, as shown here. This method of carriage, with separate clip, was adopted in 1935, prior to which time the clip was integral to the belt.

BELOW: The post-1937 seaman's clasp knife, pattern 301B. The knife was worn by all seamen suspended from a lanyard. It was stamped with the owner's name and number on a panel cast into the grip. It was an optional item of kit for Ratings other than seamen.

Ratings lunch break on the mess deck of an RN warship. *At left* a seaman wears a knife suspended from a lanyard around his waist.

Lanyard, Neck (Lanyard, Knife)

The lanyard was worn with Dress numbers 1, 2, 6 and 7 by all men wearing Class II uniform. When undertaking arms drill other than on ceremonial occasions the lanyard could be dispensed with. The wearing of the knife on the lanyard in 1, 2, 6 and 7 Dress was optional, but it was to be worn on the lanyard at all times when in working rig. When worn correctly with the jumper, the knot of the lanyard was to be positioned 3in above the tapes, with the bight of the lanyard under the silk, but not hanging too low. During the war it was common practice for the lanyard to be omitted from the uniform, despite repeated orders to the contrary. Wartime uniform changes also required that the lanyard be worn with the No. 5 Dress when it was worn with the blue jean collar.

Ditty Box

The small wooden box in which sailors kept their few personal possessions had been in use for many years. In the 1870s it was listed as the commodities box, but the term 'ditty box' was in common use. The lockable box measured 12 × 8 × 6in deep. In 1937 it was decided that the attaché case should replace the ditty box. New entrants to the service were in future to be supplied with an attaché case once existing stocks of the ditty box were exhausted. Men in possession of the ditty box were authorized the purchase of an attaché case at their own expense.

Two Class II Ratings on short leave while visiting Australia. Both men carry the attaché case.

Case, Attaché

The attaché case was first available as an optional item for all Class I, II and III Ratings in 1935. By 1938 it was authorized as an alternative to the ditty box for Class II Ratings, and by 1939 issue of the case had replaced the ditty box. Made of compressed fibre with a brown finish, it measured 14 × 10 × 5in deep. As well as for storage of personal items the case made a good weekend bag.

Case, Suit and Kitbag

The kitbag has long been associated with sailors. While it was suited to the storage of Class II uniform, it was not well suited to storing the fore-and-aft rig of Class I and Class III Ratings. Ratings in these classes were authorized the use of a suitcase. Prior to 1921 the case had not been a standardized item, but AFO 3202 issued in September 1921 introduced a standard Admiralty-pattern suitcase in two sizes. The large size case measured 26 × 15 × 6in and was for issue to CPOs. For other Class III Ratings a 20 × 15 × 6in suitcase was provided. The use of the larger case was later extended to Class III Ratings where storage facilities for it existed on board. The cases were made of compressed fibre covered in green Willesden canvas.

In 1937 requests were made for the suitcases to be made deeper. It was found that increasing the depth by no more than 1in would still allow the case to be fitted into the on-board storage, so the increase in size was approved. The suitcases could be purchased by Class II Ratings for use ashore or in barracks. The suitcases were not allowed to be used on board any ship by seamen Ratings who were only authorized the use of the kitbag.

The 12 × 8 × 6in wooden 'Ditty Box' was used for the storage of a Rating's personal effects since the mid-1800s. In 1937 it was finally replaced by a small 14 × 10 × 5in attaché case that doubled as a weekend bag on short leave.

Admiralty-pattern green canvas suitcase. This is the large-size case 26 × 15 × 7in for issue to CPOs. It was an optional item for other Class III Ratings, who were otherwise provided with a 20 × 15 × 7in case.

LEFT: *At left* is a small size kitbag in white. It has been personalized with an ensign and anchor motive and the owner's details. *At right* is a linen bag for the storage of clean underwear, but also used as a dirty washing bag. The linen bag has been adorned with the crest of HMS *Ajax*.

By World War II the decorative ropework on the kitbag's base had been replaced by a simple, full-width grab strap inset into the side seam at each edge of the base, as shown on this mid-war production bag. A lockable bar fastener and padlock for use with the kitbag, shown at upper left, was introduced in February 1922. Fitted at the mouth of the bag, it provided a deterrent to theft of contents during storage or while in transit.

At left is a pre-World War II kitbag with a rope-edged section on the base for stamping the owner's name. World War I-production kitbags had a brass plate fitted within the ropework, but it was discarded as an economy measure. *At right* is a smaller size kitbag in black. During World War II white kitbags were often dyed black to help prevent aerial observation of mass troop movements that might be indicated by piles of kitbags on railway platforms, docks and quaysides.

A wartime production kitbag, with full-width base strap, being carried aboard a US lend-lease destroyer, Canada, 1940.

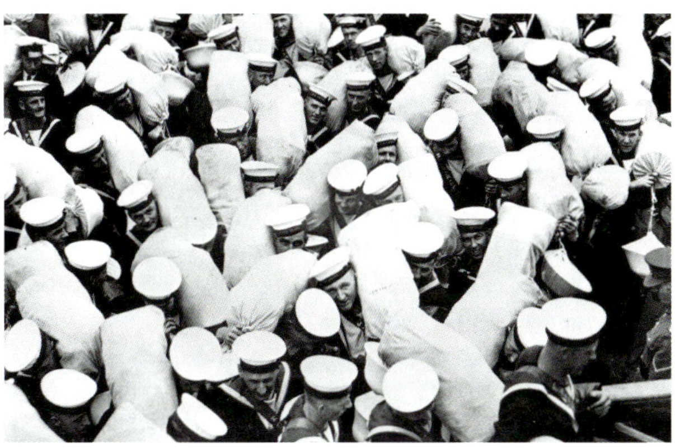

Seamen Ratings boarding a ship laden with small kitbags and Sun Helmets in calico covers. All the men also carry gas masks over their shoulder.

Bag, Travelling

In May 1945 the Bag, Travelling was added to service stocks as an alternative to the suitcase. The canvas holdall measured 25 × 13.5 × 6.5in, and was soon to prove a popular item. Typically the 'Pusser's grip' was only for issue to Officers and Class I and III Ratings, but as supplies met demand sale was eventually extended to Class II Ratings. In August 1945 a special padlock was made available for use with the travel bag.

The 'Bag, Travelling', better known as the 'Pusser's Grip'. The bag was introduced in May 1945 for optional purchase by Officers and Class I and III Ratings.

Identity discs. *At left* a disc issued to a Leading Radio Mechanic (LRM) of the Fleet Air Arm; the reverse shows the blood group. Centre is an Officer's tag, without number and a blank reverse. *At right* a Rating's tag with name and religion on the obverse, and rank and number on the reverse. This tag is suspended from a shortened Bosun's Call chain, along with a crucifix.

A silver identity bracelet worn by a Sub Lieutenant of the RNVR. S/Lt Turney was attached to the Combined Operations base HMS *Copra* for beach duties.

Identity Discs (ID tags)

Prior to March 1939 identity discs were held in bulk by accountant Officers and not issued to individuals. CAFO 805 of March 1939 stated that a single red fibre identity disc was in future to be issued to all naval personnel, stamped using ⅛in or 5/32in letters and figures, with the name, branch, official number and religion of the Rating to whom they were issued. Officers' discs were stamped with the name, rank and religion. The reverse of the disc was to be stamped with the blood group in both the International (Landsteiner) and the Moss Notations, namely AB and I, A and II, B and III or O and IV. In peace time the discs were to be wired to the respirator container, but at the outbreak of war they were to be removed and worn around the neck on a length of cord. It was common for discs to be worn on chains privately purchased for the purpose, or added to neck chains already worn.

In May 1940 it was announced that identity discs bearing the abbreviation for submarine detector (SD) Ratings were considered to be a major security risk and might lead to extensive interrogation should they be captured. Because of this risk all ID discs bearing the SD abbreviation were to be destroyed and replaced with plain discs not showing the Ratings' non-substantive branch.

In December 1941 it was announced that a report had shown that casualties were found to be rarely wearing identity discs. The importance of wearing ID discs was stressed, and a reminder of the instructions regarding the discs impressed on all concerned. The following year AFO 4978 of October 1942 again raised concerns over the continued laxity regarding the wearing of ID discs. The *Manual of Victualling*, BR93 volume I, gave precise details on the wearing of discs. However, despite this and special steps to ensure that both Officers and men wore their discs at all times, great difficulty continued to be experienced in the identification of casualties who had not been wearing their ID discs.

Pay Book S.43A

The *Sailor's Pay Book S.43A* was issued to all naval and marine Officers and men including the WRNS serving on shore in the United Kingdom or afloat in home waters. Ratings serving abroad were not provided with pay books, but they were required to have an identity card. In 1940 the *Sailor's Pay Book S.43A* and the separate sailor's identity card were combined to become the *Sailor's Pay Book and Identity Book S.43A*. All the old *Sailor's Pay Books S.43A* were withdrawn and replaced by the new book. The name of the *Sailor's Pay Book and Identity Book* was changed to *Naval Pay and Identity Book* in 1942; it retained the number S.43A.

Page one of the book showed details including the full name and rank if an Officer, or the full name, date of birth and official number if a Rating. The last page contained a photograph if a Rating. Officers were not required to enter their date of birth or have a photo as they also carried a separate naval identity card. The pay book was to be carried on the person at all times by Ratings.

In July 1945 a simple waterproof protective 'paybook wallet' was introduced for use with the *Naval Pay and Identity Book*. A gratuitous issue was made to Ratings serving on stations where tropical clothing was worn, and to Naval Air Ratings serving on first line squadrons. The wallet was made of waterproof fabric with a fold-over top flap, which helped to prevent damage to the pay book from water or perspiration. The wallet was of the same pattern issued to Army ranks for jungle operations.

Examples of S.43A *Sailor's Pay Book and Identity Book*, later renamed *Naval Pay and Identity Book*. At *bottom right* is the protective wallet introduced in 1945.

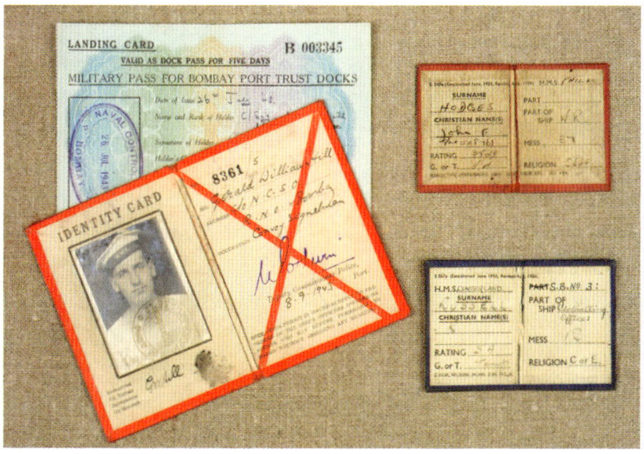

At left is an identity card and dockyard pass issued to a convoy signalman. *At right* are examples of 'station cards'.

Station Cards

Station cards were coloured red for port watch, green for starboard and blue for special duties; the cards showed a Rating's details and place of duty. Also shown was a man's entitlement to a rum issue, the card being marked 'UA' for underage, 'G' for grog and 'Temp' for temperance. The cards were handed to the Watch Office or Quartermaster when leaving the ship or establishment, providing a record of which Ratings were ashore. Defaulters (men under punishment) would forfeit their cards, and anyone not back on board at the expiration of leave would find their card being held by the Master-at-Arms pending a satisfactory explanation of absence or the results of disciplinary action.

Type, Name

The name type was issued to Ratings to mark their name on kit. It consisted of a block of $\frac{3}{8}$in reversed wooden letters that when coated in paint and stamped on a garment left the owner's name indelibly marked. Regulations stated that blue clothing was to be marked in white paint and white clothing in black paint or marking ink. Jerseys, comforters and similar items could be embroidered in red worsted or on a white tape sewn over with blue worsted.

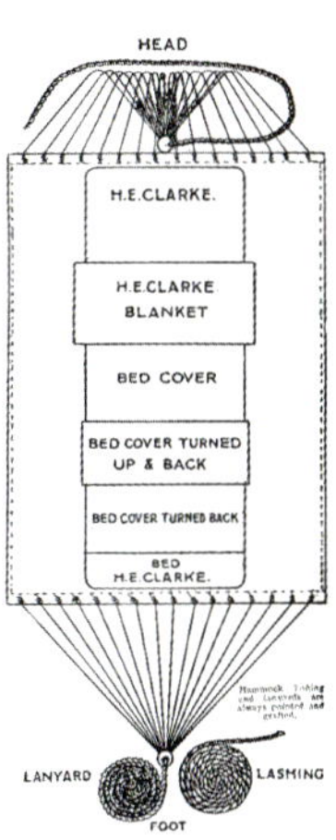

An illustration of a hammock and bedding, showing the position of ownership markings.

Naval regulations required that all clothing was marked with the owner's name. A wooden name type was issued to Ratings for this purpose, as shown *above*. *Below* is the large size marking applied to hammocks and kitbags. The large type blocks were loaned as required and were not personal issue.

Overcoats, jackets and oilskins were to be marked across the shoulders on the inside. Trousers, drawers and cholera belts were to be marked on the inside of the waistband at the back. Jumpers were to be marked on the back, close up to the collar seam. Canvas jackets and overalls were to be marked close up to the collar seam. Flannels were to be marked on the tail, caps in the crown, cap covers on the band inside, towels at both ends horizontally, scarves diagonally across the corner, boots and shoes inside the upper, and jerseys on the inside of the bottom hem. The marking of Class I and III was to conform as closely as possible to that laid down for Class II.

When kit musters were ordered, clothing was to be displayed in such a manner that the name was clearly visible.

Declared Clothing Stamp

Uniform Regulations (Ratings), Section 1, Sub-section 6, gave details of the Declared Clothing stamp that was used to mark all declared clothing. First promulgated in AFO 1402/33, the stamp consisted of the entwined letters DC within a circle. The marking was stamped over the original name marking when any clothing was transferred from one man to another. The stamp was to show that the garment had been transferred legally and not stolen or otherwise misappropriated. It was held by the ship's Accounting Officer or designated Officer for use as required.

A faint Declared Clothing stamp marked over the original owner's name on a post-1943 'Jumper, Serge' with bound and reinforced collar. The DC-identified clothing transferred from man to man with official approval.

Diagram EFO 65/34 showing the design of the Declared Clothing stamp, as promulgated in AFO 1402/33. Stamps were made on board, and thus can show a degree of variance.

All naval ratings were well versed in the art of 'make and mend'. The housewife – or 'hussif' as it was commonly called – was essential in the maintenance of uniform clothing. *Above* is a rolled wartime economy hussif with single tape, *below* is an open early war pattern with double-tape closure and buttoned pockets.

Housewife (Hussif)

The housewife was issued to boys in harbour training establishments as part of their compulsory kit. Outside training establishments the housewife was not issued, but it was an optional item of kit for all Class II Ratings.

The standard housewife was a blue cloth roll approximately 19 × 5.5in. It had one small open top pocket with four flannel needle swatches, and two flap top pockets with button closure and a double closure tape. A wartime economy housewife was also produced. The austerity version was 14 × 5.5in. It had a single, open-top pocket with two flannel needle swatches, a single fold top pocket and a single closure tape.

Housewife contents generally included needles, pins, cotton and thimble. Buttons and other items were added as needed and to personal requirements.

Respirator, Service

The naval service respirator was of the standard military pattern but with a long hose that allowed it to be worn with the haversack and canister to the small of the wearer's back. This method of wear required an additional waistbelt to hold the haversack in position. It was considered that wearing the haversack in this manner was better suited to naval requirements.

In January 1939 moves were made to equip all personnel under training with respirators. Stocks of respirators were to be held in bulk by all shore training establishments, but only to be issued on the outbreak of war. The standard service respirator was to be provided to youths, with Cadets and boys receiving the civilian duty respirator. In July 1939 retired and reserve Officers called up for service at the Admiralty or outlying shore duties were ordered to retain their civilian respirators, with no issue of the service respirator being made. Officers moving to subsequent appointments where the service respirator was required, including all ships and naval shore establishments, were to exchange their civilian respirator for the service pattern.

The production of respirators was modified in 1941 when some components previously made of non-magnetic materials were changed to mild steel as an austerity measure. The components affected were the valve holders (aluminium), valve guards (brass) and eyepieces (stainless steel). The change reduced the unnecessary use of valuable metals, but introduced a problem with regard to magnetic compasses because the mild steel components were magnetic and thus had a detrimental effect on compasses. Thus Officers taking bearings whilst wearing or carrying a respirator with mild steel components would have inaccurate readings. All Officers who were required to take bearings or work in close proximity to magnetic compasses were to equip themselves with respirators having no magnetic face-piece components. Every effort was made to ensure that Executive Branch Officers were equipped with non-magnetic respirators, and any Officer found to be in possession of a respirator with magnetic components and who worked near a magnetic compass was required to ensure that it was changed.

A Royal Navy-issue respirator and haversack. Similar to the army type, the naval respirator had a long hose section and the haversack had a waist strap allowing the haversack and canister to be worn at the back when in use, preventing entanglement or obstruction in the confines of a turret or below decks. Early production face-pieces had a fabric covering, with later examples being plain rubber, as here.

Heavy AA gun crew from HMS *Rodney* wear respirators and anti-gas clothing, *c.*1939. The long hose and waiststrap of the naval pattern mask and haversack are clearly evident.

Royal Navy issue respirator haversack straps. *Above* is the waist strap used with long-hose respirator haversacks. *Below* is a shoulder strap with typical RN name stamping.

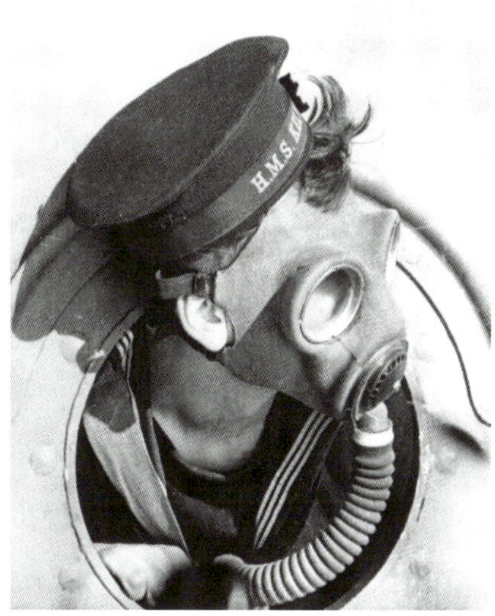

RIGHT: A Class II Seaman Rating peers through a porthole while wearing a gas mask with fabric-covered facepiece.

LEFT: Watch Coat worn with the anti-gas respirator in the alert position on the chest. When worn at the alert the respirator was quickly and easily donned.

Anti-gas respirator haversacks with waist straps are carried by men following the presentation of distinguished service medals, March 1940. The rating fourth from the left carries the pre-war pattern of haversack.

From the outbreak of war, respirators were to be carried at all times. This order was relaxed in August 1942, when respirators were only to be carried in field service marching order, or when mounting guard (other than ceremonial), or when ordered by the Flag or Commanding Officers.

The respirator haversack was not to be scrubbed or cleaned with Blanco. It was permitted to use Mill's cleaner No. 700 or Pickering's khaki equipment cleaner, as these did not damage the waterproofing.

Davis Submarine Escape Apparatus (DSEA)

The Davis Submarine Escape Apparatus (DSEA), also known as the Submerged Escape Apparatus, was adopted by the RN under AFO 1096/30. This oxygen re-breather set allowed crewmen from stricken submarines to escape from a depth as deep as 330ft using a 15oz canister of CO_2 absorbent and a 2cu ft oxygen cylinder. The apparatus also contained two small steel 'oxylet' capsules containing a supplementary supply of oxygen. The DSEA also provided buoyancy for the wearer once on the surface. The following numbers were issued in 1930: HMS *Dolphin* and 5th Flotilla, 1,030 sets; HMS *Lucia* and 2nd Flotilla, 310 sets; HMS *Vulcan* and 6th Flotilla, 230 sets; HMS *Cyclops* and 1st Flotilla 420 sets; HMS *Medway* and 3rd Flotilla, 760 sets.

The Davis Submarine Escape Apparatus (DSEA), introduced into RN submarine service in April 1930. THE HISTORICAL DIVING SOCIETY

Crewmen from a Chariot (Human Torpedo) wearing diving suits and oxygen re-breather sets based on the Davis EBA.

The Action Ration

The supply of an iron ration for men closed up at action stations was first given consideration during 1942. It had been found that men under the extreme stress of continuous air or surface actions were disinclined to eat. While at action stations men were fed at their posts, often being provided sandwiches, stew or whatever food could be easily prepared and delivered. This 'action messing' was supplemented by liberal quantities of fluids to help prevent dehydration: cold weak tea with sugar but no milk, oatmeal water, sweetened lime juice with water, and tinned fruit in juice were standard.

By the spring of 1943 a supplementary ration for use during action stations had been developed by the Admiralty. The 'Action Ration' was packaged in a small – 3¼ × 1¾ × ⅝in – airtight tin. It contained six Horlicks' tablets, four barley sugar tablets and one packet of chewing gum. The ration was considered sufficient for a single day, with ships carrying enough rations for three days' supply. It was stressed that the ration was designed to supplement daily food and was not to replace it. It could only be issued when the circumstances necessitated its use, and required certification by commanding Officers.

The supply of Action Rations was restricted to battleships, aircraft carriers, cruisers, destroyers, landing ships, and vessels employed on long-distance escort work.

Action Rations were a supplementary food source for use during action stations. The small tin contained six Horlicks' tablets, four barley sugar tablets and one packet of chewing gum.

Men on a destroyer's mess deck sit down to their evening meal, after which tables would be cleared away and hammocks slung. The man already in his hammock would probably have been a 'watchkeeper' resting between his watches.

5 Badges

RNVR and RNR Sleeve Badges

In 1924 the old R (Royal Naval Reserve) and V (Royal Naval Volunteer Reserve) sleeve badges were discontinued. AFO 871/24 substituted new badges bearing the complete designation RNR and RNVR respectively. The old badges were not to be used up, but were to be returned to stores. Economy X pattern woven badges were produced from 1941.

Authorized sleeve letters included RNVR (Royal Naval Volunteer Reserve), RNR (Royal Naval Reserve), RNSR (Royal Naval Supplementary Reserve and Royal Naval Special Reserve), SWS (Shore Wireless Service), RNVWR (Royal Naval Volunteer Wireless Reserve), and SBR (Sick Berth Reserve). A number of unauthorized badges were commercially produced, including Royal Naval Patrol Service badges bearing the lettering RNP, RNPS and RNRSR. In 1943 Naval Artificer Apprentices serving with the Fleet Air Arm were found to be wearing the letters FAA in gold. Ratings wearing unauthorized badges faced disciplinary action.

Substantive, Non-Substantive and Miscellaneous Badges

The substantive (rank) and non-substantive (trade) badges worn by Royal Navy Ratings during World War II are a study in their own right. Examples of many badge types are illustrated throughout this work to provide an overview of the various patterns in use. Numerous changes to non-substantive badges took place in the period 1930–45, with old badges being modernized and many new badges added as the trade groups were expanded to meet the needs of a rapidly changing Navy.

One of several new branches formed during the inter-war years was that of the photographer Rating. Photographers wore the badge of their parent branch (Gunlayer or Torpedoman) while employed as photographers, which often resulted in their employment being misunderstood. To rectify this, two new badges were authorized in November 1930: Photographer 1st Class, showing a camera with star above, and Photographer 2nd Class, a camera without star. The badges were to be worn on the right cuff in addition to the Rating's usual non-substantive badge. The wearing of badges by CPOs was optional, but they were to be worn on the collar instead of gunnery or torpedo badges.

In June 1937 the previously non-substantive photographer's branch became a separate substantive branch in its own right. New Ratings and non-substantive badges were introduced: CPO and PO Photographer, a camera with crown above; Leading Photographer, a camera with a star above; and Photographer, a camera.

Prior to 1931, supply and writer Ratings had worn a plain gold star, and cooks a silver star. Officers' stewards and Officers' cooks wore a plain gold disc. In October 1931 the badges of accounting branch Ratings were replaced by a star encircling the letters 'OS' for Officers' steward, 'OC' for Officers' cook, 'C' for cook, 'W' for writer and 'S' supply rating. The old pattern badges were to be issued while stocks remained. The Officers' cook branch had been amalgamated with the ships' cook branch in July 1931. All new entrant cooks were to serve as naval cooks, with the old ships' cook and Officers' cook branches being allowed to die out through natural wastage. Unlike other trade groups, the accountant non-substantive badges used during World War II did not bear crowns or stars to identify levels of training. The use of a crown and stars was not authorized until post war.

In July 1938 the gunnery branch was reorganized to take account of the increasing complexity of weapons and the rapid development of anti-aircraft gunnery. In future gunners were to specialize in a particular field of gunnery. Four gunnery specializations were introduced, with Ratings in each group being trained to a reduced level of general understanding of gunnery, but with an increased knowledge of their particular specialization. The new specializations for gunners were Quarters section for gun numbers and men in charge of guns (except close range AA), magazines and shell rooms, Layer section for gun and director laying and training (except close-range AA), Control section for range finding and control of fire (except AA), and Anti Aircraft section for AA lookouts, loading, laying and training of

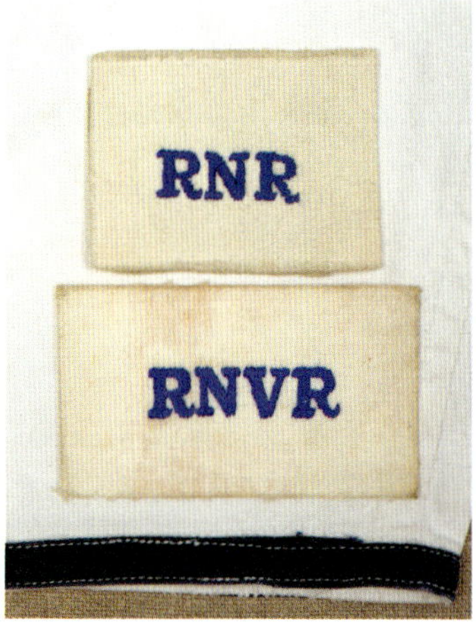

FAR LEFT: Examples of the cuff badges worn by members of the RNR and RNVR. *Top* is the pattern 268B and *below* the 269B. Badges in red and gold on blue were also produced, as were woven X patterns in blue and red.

LEFT: The unofficial RNSR (Royal Naval Supplementary Reserve and Special Reserve) sleeve badge worn by a Rating pilot (*second left at rear*). This image clearly shows the position of the pattern 94 pilot's wings badge when worn by Class II Ratings.

close-range AA equipment and control of AA fire. New gunnery badges were introduced in February 1939. These consisted of crossed guns or a single gun with a combination of crown and/or stars according to qualification, similar to the old scheme but with an added letter below: Q for quarters Rating, L for layer Rating, C for control Rating and an A for anti-aircraft Rating. Individual letters were also made available for adding to old pattern badges.

Three patterns of embroidered badge were produced: gold wire on blue serge, red on blue serge, and blue on white drill. Gold wire badges were discontinued at the outbreak of war, with the suspension of the wearing of No. 1 Dress, for the duration. Gold wire badges were hand embroidered, were costly to produce and also wasteful of vital labour essential to a nation at war. Admiralty-issue badges produced during the war were only made in red or blue, although gold badges remained on sale from Slops until supplies where exhausted. Long after the outbreak of war naval outfitters continued to produce and sell gold wire badges, including new wartime patterns such as the Combined Operations badge, DEMS and Patrol Service patterns, until gold wire eventually became unobtainable. The outfitters maintained their market by continuing to produce badges in yellow silk or rayon.

In August 1940 a new pattern of badges was authorized to help overcome difficulties in obtaining the standard embroidered badges. The new badges had the design machine-woven into the background instead of being embroidered on it. The badges were given the same pattern numbers as the corresponding embroidered badges but with the prefix X. The first batch of nine badges consisted of X1A PO red, X1B PO blue, X3A Leading Rate red, X3B Leading Rate blue, X6A PO Stoker red, X26A Leading Stoker red, X26B Leading Stoker blue, X27A Stoker 2nd Class red and X27B Stoker 2nd Class blue. A further twenty-six X-pattern badges were added in November 1940, with most badge patterns eventually being produced in the woven X pattern. The production of woven badges had been discontinued by 1949. Badges were all given Admiralty pattern numbers – the numbers had no suffix for gold badges, an 'A' suffix for red badges and a 'B' suffix for blue badges. Woven badges in red and blue had an 'X' prefix.

During the immediate post-war years only red and blue badges were issued as an economy measure, and because of extreme shortages of manpower and materials. Trials were conducted to find alternatives to gold wire, including the limited production of badges in gold nylon thread. In August 1951 hand-embroidered, gold wire badges were reintroduced on a limited scale for CPOs and confirmed POs serving overseas or on sea-going ships. As production caught up with demand, issues to other Ratings eventually followed.

ABOVE: Examples of Admiralty-pattern badges with sealing tags. *Top left* is a pattern 136 Supply Rating badge dated 1931 with wax seal. *Top right* is a 210A Radar Plot Rating 1st Class badge dated 1945. *Bottom left* is a pattern 11B Visual Signalman 2nd Class CPO cuff badge dated 1941. *Bottom right* are paired pattern 7A Visual Signalman 1st Class CPO collar badges dated 1942. Pattern badges provide excellent dated sources for determining general construction of badges, thread types and backings.

Badges and designations as extant in 1937. Non-substantive badges underwent considerable changes between 1935 and 1945, with new badge patterns and classes being added and some designations being changed, thus this illustration is only representative of the patterns. 1, Gunner's mate. 2, Director Layer. 3, Gunlayer. 4, Captain of the gun 1st Class. 5, Seaman Gunner. 6, Rangetaker 1st Class. 7, Rangetaker 2nd Class. 8, Torpedo Gunner's Mate. 9, Torpedo Coxwain. 10, Leading Torpedoman. 11, Seaman Torpedoman. 12, Leading Seaman. 13, Petty Officer. 14, Diver. 15, Visual Signalman 1st Class. 16, Visual Signalman 2nd Class (CPO and PO). 17, Visual Signalman 2nd Class (other Ratings). 18, Visual Signalman 3rd Class. 19, Trained Operator (Visual Signals). 20, Signalman (not trained operator) Visual Signals. 21, Master-at-Arms. 22, Wireless Telegraphist 1st Class. 23, Wireless Telegraphist 2nd Class (CPO and PO). 24, Wireless Telegraphist 2nd Class (other Ratings). 25, Wireless Telegraphist 3rd Class. 26, Trained Operator Wireless Telegraphy. 27, Telegraphist (not trained operator) Wireless Telegraphy. 28, Regulating PO. 29, Physical and Recreations Training Instructor 1st Class. 30, Physical and Recreations Training Instructor 2nd Class. 31, Submarine Detector Instructor. 32, Submarine Detector 1st Class. 33, Submarine Detector 2nd Class. 34, Submarine Detector Operator. 35, Photographic Rating 1st Class. 36, Stoker PO. 37, Stoker 1st Class. 38, Stoker 2nd Class. 39, Observer's Mate. 40, Acting Observer's Mate. 41, Air Gunner. 42, Telegraphist Air Gunner. 43, Mechanician. 44, Chief Armourer. 45, Chief Shipwright. 46, CPO Artisan. 47, Shipwrights and Artisans. 48, Sick Berth Rating. 49, Writer. 50, Supply Rating. 51, Cook. 52, Officer's Steward. 53, Bugler. 54, Good Shooting Badge. 55, Chief Petty Officer's cap badge. 56, Petty Officer's cap badge.

Examples of modified non-substantive badges. *At left* a pattern 19, gold wire Trained Operator (Visual Signals) badge. The badge has been upgraded to a pattern 17 Visual Signalman 2nd Class badge by the addition of a spare star pattern 46 below the flags. *At right* is a pre-1939 111A badge for a Gunlayer upgraded to the new 1939 gunnery badge system Layer Rating 2nd Class by the addition of an X117A woven letter 'L' below the guns.

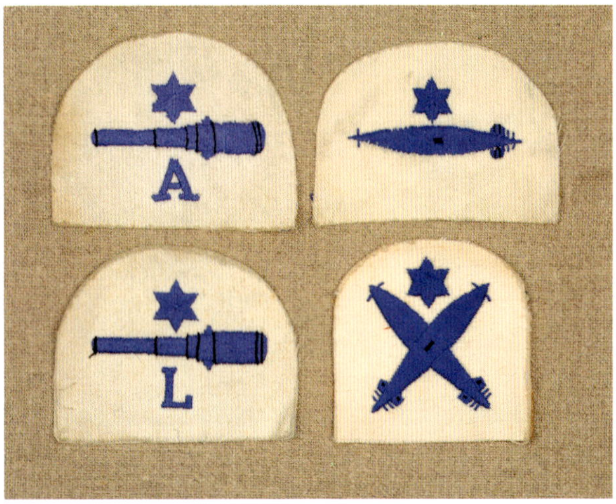

Blue badges as worn on white clothing and coveralls. *Top left* is a post-1939 163B Anti-Aircraft Rating 3rd Class badge. *At right* is the 134B Seaman Torpedoman badge, below left a post-1939 151B Layer Rating 3rd Class badge, and at right a 125B Leading Torpedoman badge.

A Class II Rating from HMS *Indefatigable* wearing the propeller over an A badge of a Naval Air Mechanic Airframes.

The reverse of typical non-substantive badges produced prior to 1945. Blue and gold badges were usually backed in hessian or cotton, and all had rear face embroidery the same colour as the front. *Top row*, left to right, a pattern 331 Air Fitter badge converted to a pattern 329 Air Fitter Ordnance by the addition of an embroidered O in yellow thread, 161A Anti-Aircraft rating 2nd Class, 134B Seaman Torpedoman. *Bottom left*, pattern 27 Stoker 2nd Class, pattern 71 bugler's badge, and *at right* the 125B Leading Torpedoman badge.

Examples of the unpopular woven badges introduced progressively from 1940 as an economy measure. The X-pattern woven badges supplemented the embroidered patterns already in service. *Top left* is a pattern X1B PO's badge, and at right a X1A badge. *Below left* the X20B Signalman (not trained operator) Visual Signals, and *right* the X64A Telegraphist (not trained operator) Wireless Telegraphy badge.

A Class II Rating from HMS *Sovereign* wearing the Pattern 115B Seaman Gunner badge on his white drill No. 6 Dress, with medals, during a visit to Malta in 1934.

A pre-October 1931 pattern 40B blue star for wear by writer, supply and cook Ratings on the White Drill Tunic.

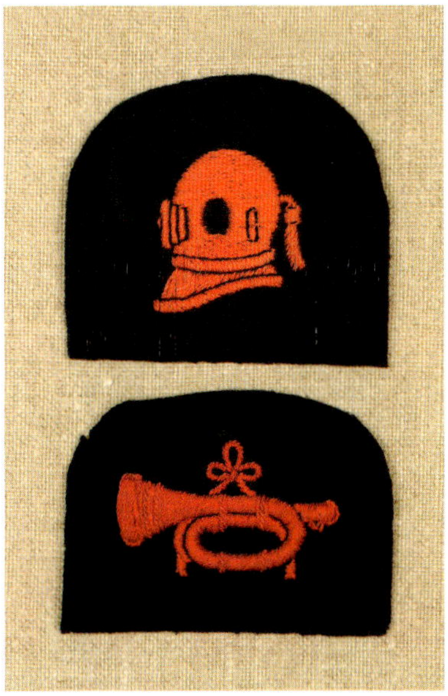

The red-on-blue diver's badge, pattern 251A. The badge was worn by all Rates without additional adornments. *Below* is the red-on-blue pattern 71 bugler's badge worn by boys and non-specialists on the right cuff.

Combined Operations Badges

All members of Combined Operations, including Royal Marines, Army, RAF and attached American and Allied ranks, were entitled to wear the CO badge that had been introduced in September 1942 (AFO 4366/42). The badge consisted of an RAF eagle and a Thompson machine gun imposed over an anchor. For Navy ranks the badge was provided in three patterns: 292A (red for wear on the lower right cuff of blue uniforms), 292B (blue for wear on the lower right cuff of white uniforms) and 291A (paired red for wear on both sleeves of the Battledress uniform). Interestingly both 292A and 292B badges were worn with the gun facing to the rear, but the paired 291A badges were worn with the gun facing to the front.

The use of gold CO badges was expressly forbidden by orders, and purchase of badges other than from official sources was prohibited. However, commercial versions in gold wire and later yellow silk and rayon, were readily available from trade sources. Signallers attached to Beach Signals Section were issued blue and white beach signals

The 292A Combined Operations badge worn on the lower right cuff by a Class II Rating.

A silk version of the Combined Operations sleeve badge. Despite RN orders that gold CO badges were not to be worn, they were available commercially. A shortage of gold wire eventually resulted in the substitution of yellow silk or rayon for gold wire on all commercial badge types. BADGE COURTESY STEVE BRITTON

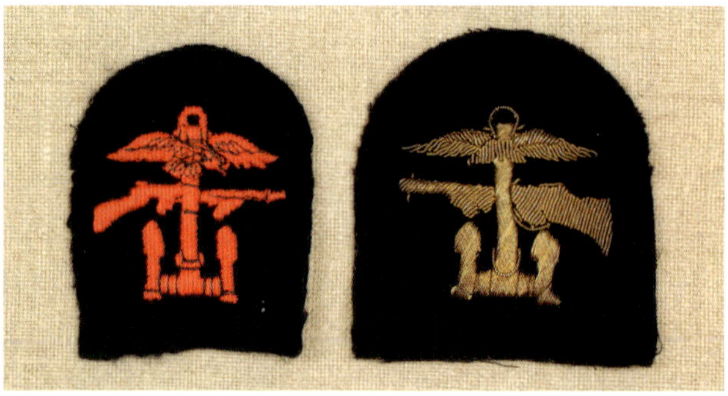

Combined Operations sleeve badges. *At left* is an issue 291A red badge – this is the right side of paired badges worn on the upper sleeves of Battledress with the gun facing forwards. *At right* is a commercial gold wire version worn on the lower right cuff of the uniform with the gun facing the rear.

flashes to be worn below the CO badge. Beach signals flashes were identical to the Arm of Service stripes worn by the Royal Signals and were obtained from Army stocks.

As well as the Combined Operations badge, special Combined Operations shoulder flashes were also worn on battledress. Those men qualified as Commandos were to wear the RN COMMANDO shoulder flashes, with non Commando-trained men using ROYAL NAVY flashes, both in white on black.

Chevrons for War Service

Authorized in 1943 War Service, chevrons first became available for issue in February 1944. The ¾in chevrons were printed red and were initially issued in sets of two, three and four chevrons; this was later extended to five chevrons. Red on drab chevrons were issued to qualifying personnel wearing BD uniform. One chevron was worn for each twelve month qualifying service subsequent to 2 September 1939. Service in qualifying organizations could be aggregated for qualification. Thus a man serving any qualifying period in, for example, the National Fire Service or Civil Defence prior to joining the RN could carry their previous service forwards to be added to any RN service.

War Service Chevrons: these were the only printed badges issued by the Admiralty during World War II.

War Service chevrons were the only printed badges issued by the RN during World War II. Non-substantive badges printed blue on white were first sealed for pattern in 1949, with stocks not becoming available from Slops until 1953.

Silver A/S-M/S Badge; Patrol Service Badge

A silver badge for members of the Royal Naval Patrol Service employed in minesweeping and anti-submarine craft in home waters was approved in May 1940.

The A/S-M/S badge consisted of a net, symbolizing the work of the fishermen who made up the ranks of the Patrol Service, and a shark, head down and pierced by a marlin spike, characterizing the shark-like activities of German U-boats, and German X-type mines. The first batch of 20,000 badges had a brooch fastening and was produced by the Royal Mint, which registered the design as No. 837751. In July 1941 issue was extended to eligible men serving on foreign stations. However, by September 1941 stocks were almost exhausted, and production of a further 10,000 badges was approved. But although the Mint was able to produce the badges, it had become impossible to obtain a supply of brooch fittings. In addition, complaints had been received to the effect that the brooch badge was easily torn from the sleeve and did not sit well. A modified design, incorporating four loops for stitching the badge to the uniform, was therefore approved. The modified badge was authorized under AFO 4261/42, with holders of the brooch pattern being allowed to exchange it for the sew-on pattern.

The badge was worn on the lower left sleeve of the uniform whilst its wearer served in auxiliary minesweepers and anti-submarine vessels, being removed from the uniform upon appointment to other duties. Men honourably discharged from the service could wear the badge on the lapel of civilian clothing. An unauthorized version displaying a stylized mine was worn by some minesweeper crews.

The silver badge for all members of the Patrol Service manning auxiliary minesweepers and anti-submarine vessels. *At left* is the first pattern badge with brooch pin fixing. *At right* is the second pattern with four loops allowing the badge to be stitched directly to the Uniform.

A privately purchased gold wire badge for an RN Patrol Service gunlayer. The Admiralty pattern 111 badges, introduced in 1942, were only made in red and blue.

A Petty Officer instructor in No. 3 Dress serving at HMS *St Vincent* in the immediate post-war period. He wears four medal ribbons on regulation cloth backing.

Typical medal ribbons for a World War II veteran. *From top left*, 1939–45 star, Atlantic Star, Africa Star, Burma Star, War Medal, Long Service and R.N. Good Conduct medal. The ribbons are mounted on a fabric swatch as per Royal Naval uniform regulations.

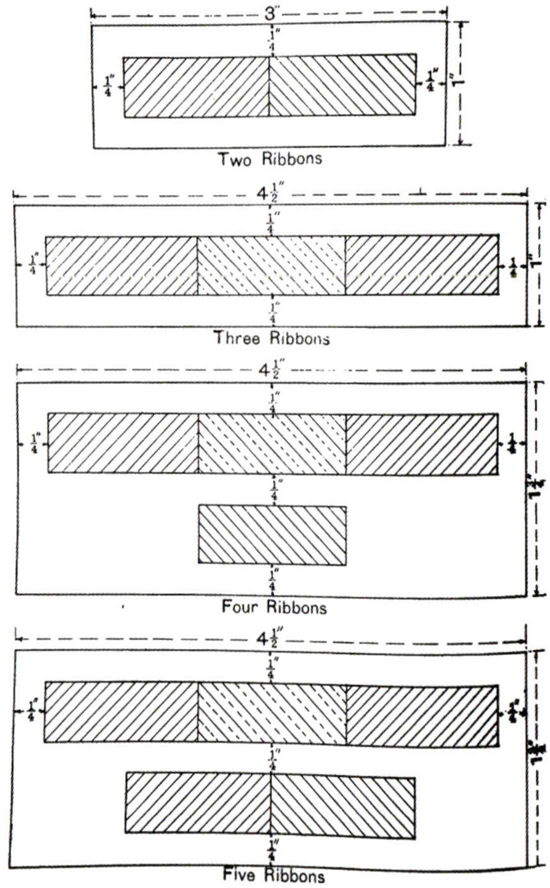

An illustration from 1933 regulations showing the method of mounting medal ribbons for wear by CPOs, POs and men, on a swatch of uniform fabric. In practice the collars of Class I and III jackets required the ribbons to be mounted no more than two wide to prevent their being hidden by the lapels.

Armlets

The Naval Patrol Armlet (Pattern 37) was worn by all POs, Seamen and Marines temporarily employed on regulating branch or patrol duties. It was worn on the lower left cuff. Early patterns of armlet were fastened with buckles, but these were replaced by press snaps as an economy measure.

Red Cross armlets were worn by naval Ratings employed in medical duties, such as first aid parties. The armlet was made of white drill fabric with a red cross imposed. Stretcher bearers wore a similar armlet with the letters SB in red. In 1944 the white armlets were replaced by khaki armlets with the red cross on a small white disc on khaki or the letters SB directly on the khaki armlet.

Armlets, Patrol, pattern 37. The armlet was worn by Ratings assigned temporarily to shore patrol duty, a much disliked task.

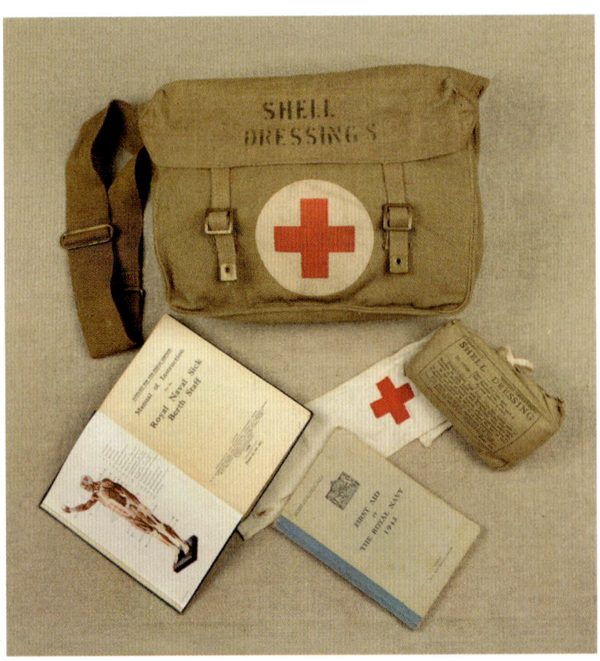

LEFT: An Admiralty shell dressings haversack of the type situated at key locations throughout the ship and carried by first-aid parties at action stations. The Red Cross armlet was worn by personnel engaged exclusively in the collection, transport and treatment of the wounded or sick. Naval Ratings serving ashore with landing parties were to wear the SB armlet. Also shown are the 'Manual of Instruction for the Royal Naval Sick Berth Staff' and 'First Aid in The Royal Navy' manual.

Abbreviations and Glossary of Terms

AFO	Admiralty Fleet Order
AWD	Action Working Dress
BD	Battledress
BR	Book of Reference
CAFO	Confidential Admiralty Fleet Order
Class I	Chief Petty Officers
Class II	Men dressed as seamen
Class III	Men other than those dressed as seamen
CO	Combined Operations
CPO	Chief Petty Officer
Chief	Chief Petty Officer
DLCO	Deck Landing Control Officer
Ditty Box	Small wooden box for personal effects
Drafted	Naval term for posting between ships or establishments.
Drill	Cotton fabric
Duck	Canvas fabric
ERA	Engine Room Artificer
FAA	Fleet Air Arm
Flannel	Woollen fabric. Class II square neck shirt
Fore-and-Aft	The uniform worn by Class I and III Ratings
HMS	His Majesty's Ship
Housewife	Sewing kit
Hussif	Sewing kit
Jean	Blue cotton fabric
Jersey	Wool knit Guernsey style garment
Jumper	Sailor's serge, drill or duck garment
KD	Khaki Drill
KUA	Kit Upkeep Allowance
MAA	Master at Arms
MGB	Motor Gun Boat
MTB	Motor Torpedo Boat
NAS	Naval Air Squadron
PO	Petty Officer
Rig	Sailor's uniform
RN	Royal Navy
RNR	Royal Navy Reserve
RNVR	Royal Navy Volunteer Reserve
Skipper	Warrant Officer RNVR and generic term for Commanding Officers
Slops	Ships cash clothing store
Square Rig	Class II seamen's uniform
Stoker	Royal Navy engineer
Tally	Ribbon bearing a ship or establishment name
TAG	Telegraphist Air Gunner
Type, name	Wooden block of letters for marking kit

The crew of a Coastal Forces gunboat scan a grey and murky sky for enemy aircraft.

Bibliography

Admiralty Fleet Orders Various issues (HMSO)
Appendix to the Navy List Various editions (HMSO)
A Seaman's Pocket Book (HMSO 1943)
Brayley *British Web Equipment of the Two World Wars* (Crowood)
Brayley & Ingram *British Women's Uniforms of World War II* (Crowood)
Brayley & Ingram *Khaki Drill and Jungle Green* (Crowood)
Brayley & Ingram *World War II British Army Uniforms European Theatre 1939–45* (Crowood)
B.R.93 *Manual of Victualling* Vols I and II (1929 and 1939)
Coleman, E.C. *Rank and Rate* Vols I & II (Crowood)
Jarrett, J.M. *British Naval Dress* (Dent and Sons Ltd 1960)
King's Regulations for His Majesty's Naval Service Various editions (HMSO)
Manual of Seamanship Vol I, (HMSO 1937)
O.U.5420 *Rate Book. Victualling and Royal Marine Stores* 1933–34, 1942
Royal Naval Handbook of Field Training (HMSO 1934)
Royal Naval Uniform Regulations 1924 & 1937 (Admiralty)
The Navy List Various editions (HMSO)

Submariners re-supply ammunition for the 4in deck gun while alongside their submarine depot ship.

Index

A Coastal Forces Motor Torpedo Boat at speed.

RELATED TITLES FROM CROWOOD

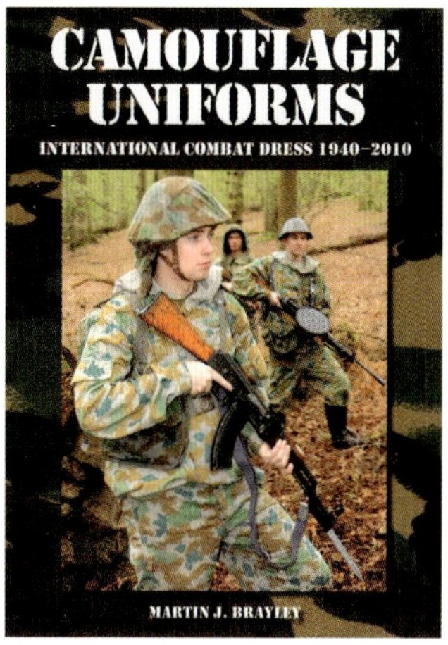

Camouflage Uniforms

MARTIN J. BRAYLEY

ISBN 978 1 84797 137 1

96pp, 400 illustrations

Khaki Drill and Jungle Green

MARTIN J. BRAYLEY & RICHARD INGRAM

ISBN 978 1 84797 109 8

144pp, 300 illustrations

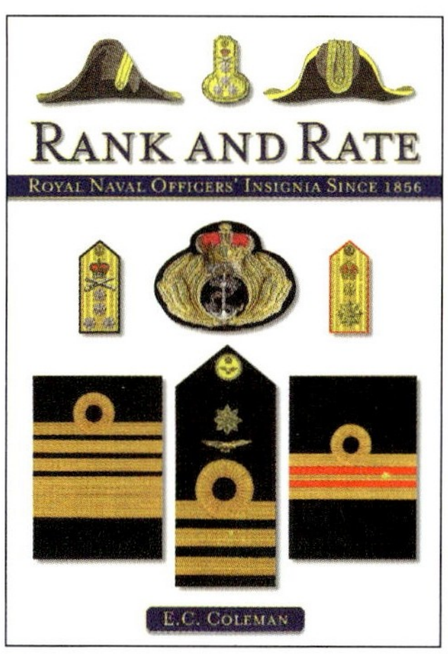

Rank and Rate Volume I
Royal Naval Officers' Insignia Since 1856

E.C. COLEMAN

ISBN 978 1 84797 138 8

96pp, 750 illustrations

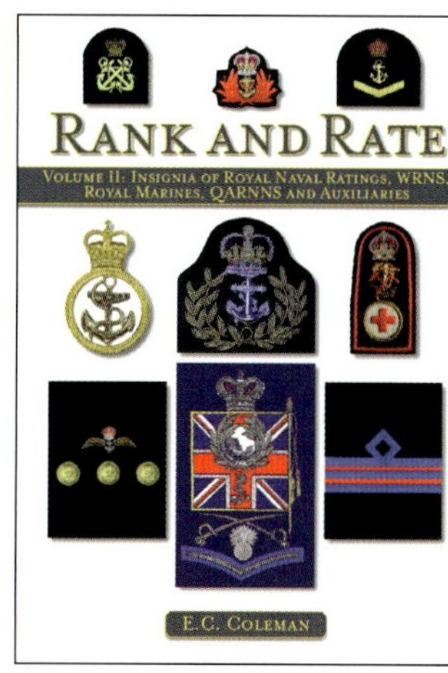

Rank and Rate Volume II
Insignia of Royal Naval Ratings, WRNS, Royal Marines, QARNNS and Auxiliaries

E.C. COLEMAN

ISBN 978 1 84797 308 5

96pp, 750 illustrations

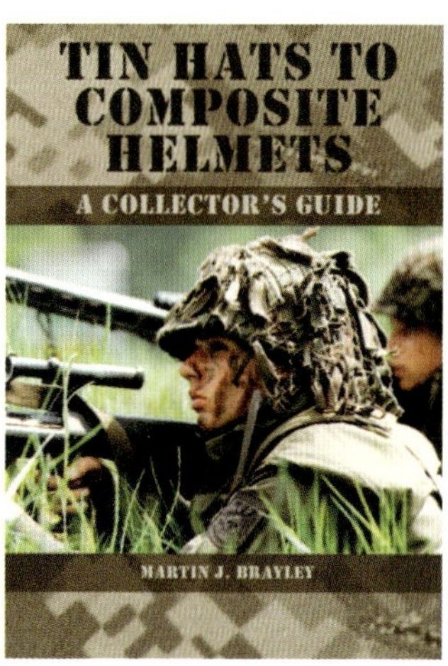

Tin Hats to Composite Helmets

MARTIN J. BRAYLEY

ISBN 978 1 84797 024 4

144pp, 700 illustrations

The World War II Tommy

MARTIN J. BRAYLEY & RICHARD INGRAM

ISBN 978 1 86126 914 0

144pp, 250 illustrations

In case of difficulty ordering, please contact the Sales Office:

The Crowood Press, Ramsbury, Wiltshire SN8 2HR UK

Tel: 44 (0) 1672 520320 enquiries@crowood.com www.crowood.com